Philosophical and Methodological Debates in Public Health

Jordi Vallverdú • Angel Puyol • Anna Estany
Editors

Philosophical and Methodological Debates in Public Health

Springer

Editors
Jordi Vallverdú
Faculty of Arts and Humanities
Universitat Autònoma de Barcelona
Barcelona, Spain

Angel Puyol
Faculty of Arts and Humanities
Universitat Autònoma de Barcelona
Barcelona, Spain

Anna Estany
Faculty of Arts and Humanities
Universitat Autònoma de Barcelona
Barcelona, Spain

ISBN 978-3-030-28625-5 ISBN 978-3-030-28626-2 (eBook)
https://doi.org/10.1007/978-3-030-28626-2

This Springer imprint is published by the registered company Springer Nature Switzerland AG.
The registered company address is: Gewerbestrasse 11, 6330 Cham, Switzerland

Preface

Public health analysis has provided a rich conceptual framework in which health determinants are intertwined with political, economic, ethical, and philosophical values. The inherent complexities of the studies about health, together with the new epistemic challenges caused by new technological forces, from e-epidemiology to statistical analysis as well as the increase of huge amounts of available data thanks to new technological possibilities such as social networks, make necessary a new, critical, and interdisciplinary study of public health. This book is the result of the debates generated by GEHUCT, an established research network devoted to the humanistic studies of science and technology specialized on public health and philosophy. Funded by the Generalitat de Catalunya, reference 2017 SGR 568, this group has led two pioneering international conferences on the philosophy of public health, held in Barcelona between 2016 and 2019. This volume is the result of several years of studies and debates among our research members as well as with international experts from the several disciplines involved: medicine, toxicology, ethics, epistemology, law, or computer sciences. Because of the continuous evolving nature of scientific research, this book not only tries to capture and explain the current problems of public health but also to face some of the close challenges which the contemporary trends are paving the way to the close future. We hope these materials will help the reader to think with more conceptual resources about that amazing and fundamental aspect of human life: health.

Arenys de Mar, Catalonia
May 2, 2019

Jordi Vallverdú

Contents

Philosophical Debates on Public Health 1
Angel Puyol

Diseases, Problems or Needs: A Framework for Public Health 9
Andreu Segura-Benedicto

Critical Epidemiology in Latin America: Roots, Philosophical and Methodological Ruptures 21
Jaime Breilh

Lost in Causality: How Epidemiology's Counterfactual Causal Inference Revolution Upholds Class, Race and Gender Inequities 47
Carles Muntaner and James R. Dunn

Coeliac Disease and Non-coeliac Gluten Sensitivity: A Clinical and Philosophical Discussion 59
F. Borrell-Carrió, E. Peguero-Rodríguez, V. Morales-Hidalgo, J. Castellote-Alonso, and A. Estany

Progress in Biomedical Sciences and Raw Data: Ethical Dilemmas 73
David Casacuberta and Simone Tassani

Design Epistemology for Public Health Research 85
A. Estany

Invisible Victims and Public Health: Epistemic Injustices in the Intergenerational Transmission of Trauma 107
María del Mar Cabezas and Carlos Pitillas

Social Determinants of Health: Citizenship Rights and Responsibilities 125
Oriol Farrés-Juste

The Public Dimension of Care: Towards a Public Ethics of Care 137
Txetxu Ausín

Genetic Prevention, Carrier Screening and the Role of Citizens: The Legitimisation of Libertarian Paternalist Policies by Deliberative Democracy 149
Baroukh Maurice Assael and Giovanni Boniolo

Public Health and Inequities Under Capitalism: Systemic Effects and Human Rights 163
Joan Benach, Juan Manuel Pericàs, Eliana Martínez-Herrera, and Mireia Bolíbar

Ethics and Public Health Policies 181
Ildefonso Hernández-Aguado, Elisa Chilet-Rosell, and María Pastor-Valero

What Is and What Is Not a Conflict of Interest in the Sphere of Public Health 195
Miguel Ángel Royo-Bordonada and Fernando García-López

Editors and Contributors

Editors

Jordi Vallverdú Faculty of Arts and Humanities, Universitat Autònoma de Barcelona, Barcelona, Spain

Angel Puyol Faculty of Arts and Humanities, Universitat Autònoma de Barcelona, Barcelona, Spain

Anna Estany Faculty of Arts and Humanities, Universitat Autònoma de Barcelona, Barcelona, Spain

Contributors

J. Castellote-Alonso Bellvitge Teaching Hospital, Bellvitge, Spain

Baroukh Maurice Assael Cystic Fibrosis Adult Center, University of Milano, Medical School, Milano, Italy

Txetxu Ausín Institute of Philosophy, CSIC (Spanish National Research Council), Madrid, Spain

Joan Benach Health Inequalities Research Group, Employment Conditions Knowledge Network (GREDS-EMCONET), Universitat Pompeu Fabra in Barcelona, Barcelona, Spain
JHU-UPF Public Policy Center at the Department of Political and Social Sciences, Universitat Pompeu Fabra in Barcelona, Barcelona, Spain
Transdisciplinary Research Group on Socioecological Transitions (GinTRANS2), Universidad Autónoma in Madrid, Madrid, Spain

Andreu Segura-Benedicto Área salud pública, Generalitat de Catalunya, Barcelona, Spain

Mireia Bolíbar Health Inequalities Research Group, Employment Conditions Knowledge Network (GREDS-EMCONET), Universitat Pompeu Fabra in Barcelona, Barcelona, Spain
JHU-UPF Public Policy Center at the Department of Political and Social Sciences, Universitat Pompeu Fabra in Barcelona, Barcelona, Spain

Giovanni Boniolo Dipartimento di Scienze Biomediche e Chirurgico Specialistiche, Università di Ferrara, Ferrara, Italy

F. Borrell-Carrió Department of Clinical Sciences, University of Barcelona, Barcelona, Spain
Gavarra EAP (Primary Healthcare Team), ICS (Catalan Regional Health Service), Barcelona, Spain

Jaime Breilh Universidad Andina Simón Bolívar, Quito, Ecuador

Mar Cabezas Centre for Ethics and Poverty Research (CEPR), University of Salzburg, Salzburg, Austria

David Casacuberta Philosophy Department, Universitat Autonoma de Barcelona, Bellaterra, Spain

Elisa Chilet-Rosell Department of Public Health, History of Science and Gynaecology, School of Medicine, University Miguel Hernández de Elche, San Juan de Alicante, Spain
Ciber de Epidemiología y Salud Pública (CIBERESP), Madrid, España

James R. Dunn McMaster University, Hamilton, ON, Canada

Anna Estany Faculty of Arts and Humanities, Philosophy Department, Universitat Autònoma de Barcelona, Barcelona, Spain

Oriol Farrés Department of Philosophy, Universitat Autònoma de Barcelona (UAB), Bellaterra, Spain

Ildefonso Hernández-Aguado Department of Public Health, History of Science and Gynaecology, School of Medicine, University Miguel Hernández de Elche, San Juan de Alicante, Spain
Ciber de Epidemiología y Salud Pública (CIBERESP), Madrid, España

Fernando García-López National Centre of Epidemiology, Institute of Health Carlos III, Madrid, Spain

Eliana Martínez-Herrera Health Inequalities Research Group, Employment Conditions Knowledge Network (GREDS-EMCONET), Universitat Pompeu Fabra in Barcelona, Barcelona, Spain
JHU-UPF Public Policy Center at the Department of Political and Social Sciences, Universitat Pompeu Fabra in Barcelona, Barcelona, Spain
Epidemiology Research Group, National School of Public Health, University of Antioquia (UdeA), Medellín, Colombia

V. Morales-Hidalgo Garraf-Penedés Paediatric Unit, ICS (Catalan Regional Health Service), Vilafranca del Penedès, Spain

Carles Muntaner University of Toronto, Toronto, ON, Canada

María Pastor-Valero Department of Public Health, History of Science and Gynaecology, School of Medicine, University Miguel Hernández de Elche, San Juan de Alicante, Spain
Ciber de Epidemiología y Salud Pública (CIBERESP), Madrid, España

Juan Manuel Pericàs Health Inequalities Research Group, Employment Conditions Knowledge Network (GREDS-EMCONET), Universitat Pompeu Fabra in Barcelona, Barcelona, Spain
JHU-UPF Public Policy Center at the Department of Political and Social Sciences, Universitat Pompeu Fabra in Barcelona, Barcelona, Spain
Transdisciplinary Research Group on Socioecological Transitions (GinTRANS2), Universidad Autónoma in Madrid, Madrid, Spain

Carlos Pitillas Facultad de Ciencias Humanas y Sociales, Departamento de Psicología, Universidad Pontificia de Comillas, Madrid, Spain

Angel Puyol Faculty of Arts and Humanities, Universitat Autònoma de Barcelona, Barcelona, Spain

E. Peguero-Rodríguez Department of Clinical Sciences, University of Barcelona, Barcelona, Spain
Castelldefels EAP (Primary Healthcare Team), ICS (Catalan Regional Health Service), Castelldefels, Spain

Miguel Àngel Royo Spanish National School of Public Health, Institute of Health Carlos III, Madrid, Spain

Simone Tassani Department of Information and Communication Technologies, Universitat Pompeu Fabra, Barcelona, Spain

Philosophical Debates on Public Health

Angel Puyol

Abstract This article introduce some of the topics of public health philosophy that are present in current debates. At the end of the text, the contributions to the collective compendium are summarized. Each of the texts presented here addresses a fundamental philosophical question related to public health. The compendium of texts is not intended to be systematic, but does comprise some of the more urgent philosophical issues in current public health. I summarize 13 chapters written by prominent philosophers and epidemiologists who share the same passion for the philosophy of public health.

Keywords Public health · Epidemiology · Public health ethics · Philosophy of medicine · Philosophy of public health

We can define public health as the efforts of a society as a whole to improve the health of the population and prevent diseases (Nuffield Council of Bioethics 2007), or, from a more political perspective, everything that we as a society do collectively to ensure conditions in which people can live healthily (Institute of Medicine 1988). Therefore, public health requires that everyone exercise a form of social responsibility in relation to health that goes beyond strictly health policies (hospitals, primary care, etc.) and includes among its primary objectives the fight against unfair health inequalities, that is, those that cannot be explained only by biology and voluntary lifestyles and can be avoided (Whitehead 1992).

This work is part of the consolidated research network "Grup d'Estudis Humanístics de Ciència i Tecnologia" (GEHUCT) ("Humanistic Studies of Science and Technology Research Group"), recognised and funded by the *Government of Catalonia*, Reference 2017 SGR 568. This work is also funded by the Spanish Government (Reference FFI2015-64858-P).

A. Puyol (✉)
Faculty of Arts and Humanities, Universitat Autònoma de Barcelona, Barcelona, Spain
e-mail: angel.puyol@uab.cat

J. Vallverdú et al. (eds.), *Philosophical and Methodological Debates in Public Health*, https://doi.org/10.1007/978-3-030-28626-2_1

Public health is a branch of medicine and its field of study is epidemiology, that is, the study of factors that determine the frequency and distribution of diseases in human populations. Traditionally, epidemiology employs conceptual tools typical of the natural sciences and epidemiologists consider themselves natural scientists seeking natural phenomena related to biological processes. The objective of epidemiology is to prevent public health problems in the population, promote good health and protect the population from diseases that affect human groups. By way of example, epidemiology intervenes when there is an epidemic, that is to say, when more significant cases of a disease are detected in a community than expected over a given timeframe (for example, an infectious disease or a sudden increase in cases of asthma, but also obesity, breast cancer, myocardial infarctions linked to tobacco use, etc.). We can say that epidemiology studies diseases that affect the community and exceed the scope of analysis and action of clinical medicine (centred basically on individual patients).

In recent years, however, epidemiology is undergoing a Copernican turn in its classical methodological approaches due to the emergence of the so-called social determinants of health (SDH). It is such an important change that the science of epidemiology is reestablishing itself to give rise to so-called social epidemiology (Berkman and Kawachi 2000), which aims to differentiate itself from classical epidemiology in that investigating the causes of diseases that affect the community should not solely involve research in biology or conditions that are external but natural to individuals (such as toxic agents), but also, and above all, in the social and non-natural conditions that surround individuals (social class, work environment, family relationships, etc.). It is therefore about looking for the (social) causes of the (natural) causes of socially prevalent diseases. Although the impact of SDH on epidemiology as a science is new (in 2008, the WHO published the Commission's Report on Social Determinants of Health, led by M. Marmot), there has been evidence of these *causes of causes* since the 1970s and 1980s (the Black Report and the Whitehall Report).

This new approach to epidemiological science has led to the emergence of numerous new philosophical questions (epistemological, ethical, ontological, among others), which have begun to concern the more aware epidemiologists (though still trained in the classic methodological concepts of the natural sciences). However, these questions have not yet been treated with sufficient rigour, precisely because scientists—in general—lack the habit and training to address them, given that it is a relatively new field of study that relates science, epistemology and ethics in a novel way.

Each of the texts presented here addresses a fundamental philosophical question related to public health. The compendium of texts is not intended to be systematic, but does comprise some of the more urgent philosophical issues in current public health. We have compiled thirteen chapters written by prominent philosophers and epidemiologists who share the same passion for the philosophy of public health.

Andreu Segura makes an ontological incursion into the idea of public health, emphasizing the need to conceptually define health and public health. His text distinguishes between two approaches: the pathogenic, aimed at curing and preventing disease, and the salutogenic, which aims to detect and promote the causes of good health. This latter perspective, which the author believes to represent the ideal in the field of public health, has the drawback of not being backed up by a widely accepted theory of health beyond the traditional theory of disease inherent in the pathogenic model. Segura notes that it is our responsibility to combine both perspectives in a balanced and consistent manner.

Jaime Breihl reviews the contribution of Latin American critical epidemiology that began in the 1970s as a counterpoint to the "knowledge illusion" of linear, reductionist causal thinking. However, limited by its subordinate position with respect to mainstream medicine, this insurgent academic position grew as a counteractive intellectual and political tradition that based its initial impulse on epistemological clarity and activism. Contrary to the linear functionalist thinking of conventional medical science—which separates the understanding of health conditions from their social and cultural contexts—critical social medicine broke away from that positivist framework. The historical roots, epistemological and methodological ruptures of critical epidemiology are discussed. The author describes how the social determination of health paradigm was formed from dialectical, complex, critical thinking with its potential for an emancipatory perspective of health reform. This work presents a complex form of thinking to decipher the real requisites of hard rigorous epidemiology, which depends not only on the accuracy, confiability and validity of its empirical methods, calculations and observations but on a rigorous understanding of society and health.

In her text, Anna Estany argues that public health research responds to the scheme of a design science. The aim of this chapter is to explore just how far design epistemology can be adopted as a methodological framework for research in the field of public health, and precisely what it adds to classic epistemology or even to the standard methodology of applied science. On the one hand, she analyses different approaches to design epistemology and related terms and expressions such as "design thinking", "design theory" and "designerly ways of knowing"; on the other, she considers where to situate public health within the field of academic knowledge and research and the disciplines involved in this field of research, ranging from biomedical science to sociology and demography. Finally, she shows how design epistemology can offer solutions to the challenges currently faced by a phenomenon as complex as public health.

A joint article by Francesc Borrell, Eva Pegueró, V. Morales-Hidalgo and Anna Estany discusses the philosophical reasons that justify the incorporation of a new disease within the public health sphere. In the case of coeliac disease and non-coeliac gluten sensitivity (NCGS), an important tension arises between a pragmatist position, in favour of diagnosing NCGS, and an opposing essentialist conception, since no "biological core" specific to NCGS can be identified. When proposals for new diseases appear, practitioners adopt positions in favour or against their acceptance, supported by different epistemological approaches. In this work, we analyse these approaches, which lie behind many nosological and taxonomic discussions in medicine.

In the following text, David Casacuberta discusses some ethical and epistemological problems of Big Data in public health issues. Casacuberta asks how we can ensure that a scientific result is sound, that no fraud has been committed and that scientific progress is guaranteed in this new era of big scientific data. In order to guarantee reproducibility, especially in the biomedical sciences, third parties need to be given access to analytical datasets (either raw or processed), the software used, the code developed by researchers, a very detailed description of the methodology and all the relevant metadata. The author shows how sharing data is mandatory in order for science to progress. Each published paper should not be seen as an untouchable revealed truth, but as an opportunity for dialogue, confrontation and verification. This can be done only if all raw data are shared and each researcher has the right tools to reproduce an experiment.

Following this, Miguel Angel Royo and Fernando José García-López discuss the ethics of conflicts of interest in research on public health. In this sphere, conflicts of interest can arise in the field of research when studies are conducted on the health effects of pharmaceuticals, chemicals, tobacco, alcohol or food and in decision-making regarding policies related to these same products. In this chapter, the authors present several examples of conflicts of interest in public health and a series of recommendations for their management beyond their public declaration, which is necessary but insufficient. Royo and García-López only speak of a conflict of interest when external secondary interests, usually of a financial nature, are opposed to the primary professional interest (preserving the health of the population in the case of public health). In addition, biases due to internal interests are bidirectional and difficult to quantify, while those deriving from financial conflicts of interest are predictable, unidirectional and large-scale. Although these are different phenomena, they are often mixed up with one another to create confusion and divert attention from financial conflicts of interest.

Mar Cabezas and Carlos Pitillas apply the concept of hermeneutical injustice to social and clinical misconstructions regarding the case of intergenerational transmission of trauma and subtle relational trauma, as well as to the lack of a child-centred perspective, which obscures not only the social study and understanding of the problem, but also the victim's ability to make sense of her own experiences from generation to generation. In order to overcome the invisibility involved in these cases and its negative impact on public health policies, the chapter argues that such cases constitute an instance of epistemic injustice. Several normative problems and paradoxes concerning the ethics of child health, such as the questions of thresholds in public health policies or conflicts in parental responsibility, are addressed through the lens of epistemic injustice. As a result, contrary to reactive views, they advocate for inclusive preventive measures as a matter of justice to avoid the perpetuation of invisible injustices.

Txetxu Ausín explores the possibility of a public ethics of care. As essentially social beings, we humans are fragile and dependent on what is outside ourselves-others, institutions and sustained and sustainable environments. Care-related obligations by the (social) State come into play here, as in the case of healthcare systems. These can be characterized as "due cares" (due actions) because the action is an expected event within a social context of values and guidelines. In this context, an omission that causes or enables harm, as in exclusion from healthcare, qualifies as an "undue inaction". In the criminal legal sense, the idea of "nonfeasance" refers to bringing about a result by not avoiding it, by violating a special duty, which equates with its causation—the prototypical example is the crime of failing to assist. Here the question not only one of intentional omission but of neglect and abandonment. To illustrate the theory, Ausín employs the example of broad sectors of the Spanish population being excluded from healthcare and medication, and not only undocumented migrants but also the poor and unemployed. The State is shamelessly neglecting important segments of the population, and causing harm—especially to the more vulnerable—through an ethically reprehensible omission of care. This case illustrates the need to define what we have called "a public ethics of care" that combines public ethics obligations with the responsibilities of care—since government is shaped by the ethics of care.

Fully addressing the issues of ethics and political philosophy in public health, Ildefonso Hernández-Aguado, Elisa Chilet-Rosell and María Pastor-Valero analyse the role of ideology in public health policies. Although the debate on whether public health research and practice should engage in political action touches on several issues, the authors focus on two main functions of public health here: governance and advocacy. The government's role in promoting health and reducing inequalities is challenged in at least three areas: the capacity of public administrations; the engagement of public-private interactions; and the use of law to improve population health. An adverse global political and ideological environment has resulted in governments losing the capacity to protect health, leading to risky interactions with transnational corporations to promote health and hampering the use of law as a public health tool. In this context, public health advocacy emerges as a key action to advance population health, although there are several visions on which type of advocacy better integrates public heath principles and values. The contribution of this philosophy will benefit public health research and practice and help public health face up to its present and future challenges.

Oriol Farrés analyses the issue of the social determinants of health from the perspective of political philosophy. As we know, social determinants of health (SDH) include all kinds of social, political and economic factors that have an influence on people's health, and they are the main drivers of inequities, that is, unjust inequalities in health. According to studies on SDH, there is a clear social gradient in health, which raises a number of ethical and political questions that we must consider carefully. From the standpoint of citizenship, the redistribution of SDH is a matter of both rights and responsibilities. Farrés states that scientific (descriptive) analysis of SDH must be viewed in relation to value-oriented (normative) research on bioethics, theories of justice, and human rights. In addition, SDH and population health are directly related to the rights and duties of citizens. Solidarity (with others) and responsibility (for oneself) are two sides of the same coin when it comes to improving public health.

Continuing with theories on citizenship, Baroukh Maurice Assael and Giovanni Boniolo recall some paradigmatic ways of implementing carrier screening policies in a community. They then move on to a proposal concerning a possible bottom-up solution based on deliberative democracy (and thus on real citizens' participation in decisional processes) aimed at legitimizing libertarian paternalist strategies concerning which genetic screenings should be done and which policies should be adopted by institutions.

In the penultimate text, Joan Benach, Juan Manuel Pericàs, Eliana Martínez-Herrera and Mireia Bolíbar reflect on the global systemic effects of capitalism on health and health inequities by posing two questions: Is it possible to understand capitalism's systemic effects as a whole? Can we know the multiplicity of all these interrelated impacts on eco-humanity and on health? The text claims that as an economic and power system capitalism impregnates the social organization of all human domains, thus permeating humanity and the many facets of the health-disease production process. The example of agrofood is given to show how the capitalist agro-industry has set an extractive, predatory, commodity-driven and inequity-generating system of food production and distribution that gives autocratic power to agrofood oligopolies competing to lead a global capitalist market, ultimately producing both health-damaging hunger and obesity. The chapter highlights the great importance that social and environmental media play in the production of health, to the extent that even behaviour and the micro psychobiological effects on specific organs, cells and neurons should be understood in the light of history, power relations, economic systems, culture and ideology. Finally, the chapter translates these reflections into a criticism of how current global political institutions frame the human right to health. The authors claim that in order to resist the systemic pressures of capitalism, democratize states and international institutions and control and regulate the large oligopolies, it is essential to rethink the hegemonic discourse of the right to health so that it becomes a priority human right intrinsically coupled to the other social, political and economic rights.

In the final text in the volume, Carles Muntaner draws attention to what he considers several features of the contemporary dominant epistemology in Epidemiology and Public Health that sustain our dominant social system: what he refers to as Racialized Patriarchal Capitalism (RPC). In particular, he argues that a mutually reinforcing relationship between Epistemology and Axiology in contemporary epidemiology and public health explains their support for RPC. More specifically, the growing interest in causation in epidemiology over the last two decades has led to a recent and vibrant debate around the counterfactual/potential outcomes approach to causal inference. Without questioning the need for causal inference investigations, he argues that debates on potential outcomes/counterfactual approaches are missing aspects of causality that are necessary for understanding the type of evidence and interventions produced by epidemiology and public health and their political implications in upholding our social system (RPC). Drawing from debates on causation in social epidemiology, he critiques recent empiricist (for example, counterfactual/potential outcomes) and constructivist/empiricist epistemologies in social epidemiology in favour of a realist approach that encompasses both ontology and epistemology. Although he selects social epidemiology as the study area to illustrate his arguments, he anticipates that his conclusions may be generalizable to some additional epidemiologic subdisciplines.

The diversity of contributions in this volume reflects some of the original work that has been carried out in recent years within the GEHUCT Research Group at the Autonomous University of Barcelona (Spain), with public funding from the Government of Catalonia (2017SGR568). Some of these works were presented at the First Barcelona Conference on Philosophy of Public Health, organized by the GEHUCT group in 2016. The second edition of this international congress was held in June 2019 to give continuity to the reflections initiated by the group on the philosophy of public health several years ago.

References

Berkman, L. F., & Kawachi, I. O. (2000). *Social epidemiology*. New York: Oxford University Press.

Institute of Medicine. (1988). *The future of public health*. Washington, DC: National Academy Press.

Nuffield Council of Bioethics. (2007). *Public health: Ethical issues*. London: Nuffield Council of Bioethics.

Whitehead, M. (1992). The concepts and principles of equity and health. *International Journal of Health Services, 22*, 429–445.

Diseases, Problems or Needs: A Framework for Public Health

Andreu Segura-Benedicto

Abstract Public health can play a pivotal role between the health care system and the whole of society when it comes to designing—and also to implement them—health care policies and, above all, general public policies with impact on health. Policies which require the involvement of many other sectors besides the health sector, since many of the determinants of health, beyond the absence of disease, are unrelated to the health sector. Prioritization often involves competition between the more traditional perspectives of health that are based on a pathogenic approach and aim to cure and prevent disease and the salutogenic approach that aims to detect and promote the causes of good health or at least of resilience. The absence of one satisfactory theory of health is a limitation of the salutogenic perspective but we don't have either a true theory of the disease, beyond the description of pathological entities. Public health, understood as what we do as a society—not just as a health system—to guarantee the health of people has the responsibility and the opportunity to combine both perspectives in a balanced way.

Keywords Health needs · Health assets · Health priorities · Public health · Health public policies

Introduction

The purposes of healthcare policies and even more so of health policies—when they exist as such—are to promote (increase, improve) and protect (maintain) health and prevent preventable disease(s), as well as to cure and, if that is not possible, at least provide relief for the ill and rehabilitate them in the event that the disease has had further consequences. And, while decrepitude remains unpreventable, provide people with a dignified death.

In order to establish priorities, which are the core element of any policies, it is necessary to acknowledge the problems to be addressed or define the needs to be

A. Segura-Benedicto (✉)
Public Health Council, Generalitat de Catalunya, Barcelona, Spain

J. Vallverdú et al. (eds.), *Philosophical and Methodological Debates in Public Health*, https://doi.org/10.1007/978-3-030-28626-2_2

met, so that these can be ordered according to the degree of priority it is decided their gravity deserves and in line with the different possible preferences and desires that comprise the ideological and political alternatives available.

The first step in designing healthcare and health policies is to assess the magnitude of such needs, often defined as problems that have an impact on personal and collective health. Next, it is important to assess how susceptible such problems are to the interventions available. That is, whether they are problems that have a solution and if so whether it is applicable in the present circumstances.

Until now, the needs for healthcare policies have been established, at best, by assessing existing health problems, usually based on the morbidity and mortality experienced by populations, which is usually estimated from mortality rates and the prevalence of diseases recognized as such in official classifications.

Diseases are, then, a key concept that not only has significance in the field of clinical and medical care, but also when it comes to deciding healthcare policies. However, by disease we understand a very heterogeneous set of experiences that patients suffer, which means their consequences can be difficult to quantify. Some situations, such as homosexuality, have fortunately ceased to be considered an disease, although it was included in the international classification of diseases until its eighth revision. Such heterogeneity justifies the use of a measure that compares the impact of different diseases, such as disability-adjusted life year. Either way, the classifications in use are not exempt from limitations when assessing the burden of disease on populations.

On the other hand, if we consider that health is not, at least not solely, the absence of disease, it would seem logical that applying the aforementioned pathogenic perspective of clinical medicine is not sufficient for health promotion, since avoiding disease does not necessarily mean increasing health. In order to formulate policy objectives and rationally establish priorities, the salutogenic perspective should also be taken into account, meaning that those factors and situations that generate health in populations and the people that comprise them should be recognized and promoted. Because it is also important to be clear about what we mean by health and how it occurs.

Although they are not literally opposing concepts, health and disease are closely related, so it is important to understand this relationship in a way that is useful for establishing priorities and designing healthcare and health policies. It is particularly important to consider whether and how the two perspectives may be complementary, because they often compete for the use of social and community resources. This is a competition that can at times be counterproductive, as has been the case during the latest socio-economic crisis (Segura 2014). Because if the ultimate causes of health problems are social, the solutions should also be social, as pointed out by Geoffrey Rose and Michel Marmot (Marmot 2005). For example, the increase in suicides associated with unemployment, rather than psychiatric assistance, requires adequate regulation of the labour market and the implementation of social support policies.

The existence of an organization capable of reconciling both approaches could therefore prove very useful in rationalizing priorities in terms of healthcare and health. Perhaps public health, understood as an institution in two dimensions, health and social, can play this balancing role. The purpose of this chapter is therefore to suggest the convenience of using health and disease criteria together when establishing the objectives of las health policies—including las sanitarias and, therefore, to propose that, along with disease, improvements in the determining factors of positive health also be considered among the needs to be met through the establishment of complementary priorities and the formulation of objectives in both fields, that of disease and also that of health.

Healthcare and Health Policies: Establishing Priorities

Of the projects carried out by governments and public administrations to meet certain needs of society, public policies stand out, and among these, along with some very general ones such as economic policies, we find other more specific ones such as employment, housing, urban planning, or health policies.

Although healthcare policies are sometimes referred to as health policies, this is usually an exaggerated description, since health, as we will see later, does not depend solely or mainly on plans, programs, projects or activities in the healthcare sector. In fact, a health policy should be interdepartmental and affect all those sectors of society and administrations whose actions (and omissions) have a relevant influence on health. The WHO has made this explicit under the name Health in All Policies (Leppo and Ollila 2013).

It is an initiative that has materialized in some countries, although still in a rather limited way, given that healthcare sectoral approaches continue to prevail and in most countries healthcare policies depend specifically on healthcare-related government departments. This is true even if some are referred to as health, as is the case with the British Department of Health and National Health Service, a name that, incidentally, Iona Heath suggested replacing with the National Sickness Service, to make it more descriptive of its real function (Heath 2007).

Regardless of the investment governments make in designing policies and regardless, even, of the models used when designing them (Benoit 2013), the process of developing healthcare policies shares a series of steps that range from recognizing the needs they are intended to meet to implementing programmes and executing activities to achieve the proposed objectives.

Whether to address some needs or others or propose one goal or another is a decision related to the different political and ideological alternatives which, in the context of a liberal democracy, are defended by the different political parties. What some deem to be necessary can seem superfluous or irrelevant to others, and the same can be said of aims. If the preferred political principle is equality, the aim of implemented policies must differ from when it is that of freedom. That is then translated into aims and of course priorities.

Establishing priorities—whether related to problems to be addressed or applicable interventions—is the most genuinely political moment—in the global sense—in the design of specific policies. This is because it includes a discretionary dimension, which depends on the ideology, beliefs and political preferences of the institutions—usually the political parties—empowered to assume the responsibility of preparing them. This discretionality is not incompatible with rationality, at least of the epidemiological kind, which means establishing priorities is more logical when the problems that healthcare policy aims to improve are first quantified. That is, the needs it aims to satisfy. Secondly, it is necessary to have an idea of how susceptible these problems are to the possibilities of corrective intervention, meaning that if the aim of a certain health policy were immortality (to save human lives definitively), we could not award it any priority, since we do not have any effective intervention for such an aim at the moment. Of course, those factors that make it easier or more difficult to implement interventions that have proved effective in solving the problem in question must also be taken into account, including the interests of powerful groups and pressure groups; the quality of democratic life; corruption, etc.

Since the classic proposals by Hanlon (1974) and the simplest contributions listed by Pinneault (Pineault and Daveluy 1986), all used by traditional health professionals, and especially as an academic exercise, more complex and sophisticated methods for establishing priorities have been developed (Goetghebeur et al. 2008), although quantifying and assessing needs is the first phase to be developed in all prioritization processes. What is it intended to improve and to what extent? What is it intended to prevent and to what extent?

Our interest therefore lies in considering whether the needs to be met have more to do with health promotion than with the prevention and control of disease. Because if health is not only the absence of disease, then the two aims need not coincide. Moreover, unnecessary and inadequate medicalization can be counterproductive in terms of health. The importance of iatrogenesis would seem to indicate the harmful effects of medical interventions, the negative influence of which has been repeatedly shown in recent times (SESPAS/OMC Iatrogenesis, n.d.; Makary and Daniel 2016).

Clarifying Some Concepts

Given society's interest in improving health and controlling disease, it is necessary to specify the meaning of both concepts. And, since health has historically been considered an disease-free state, it may make more sense to begin by asking what is meant by disease so as to then consider what we mean by health and, of course, to describe and analyze the relationship between the two concepts. Doing so will allow us to then consider how to best intervene to improve and protect health and to prevent diseases and cure, relieve or rehabilitate patients who suffer them.

Disease(s) and Health

One of the first questions posed by a clinician is what is wrong with their patient, why they came to the consultation and if possible, to diagnose the disease afflicting them. Discerning which disease or diseases are affecting the patient makes it possible for the doctor to establish a prognosis and decide on treatment. Adding together the diseases that affect all patients, or the individuals that make up a population, is therefore one way of deciding which health problems need solving and what healthcare policies are to be designed for.

Of course, if we wish to act with some degree of precision, it would be very useful to know the extent to which doctors are able to accurately diagnose the diseases that afflict patients and that will later make up morbidity statistics. There is no inconsiderable variability in the diagnosis (and also treatment) of diseases and it depends not only on professionals' ability and the quality of healthcare systems (MM1 et al. 2008; Naaktgeboren et al. 2016) but also on the frequency of diseases in a given population, given that the positive predictive value of a diagnostic test is related to the prevalence of the condition being diagnosed (Menten and Lesaffre 2015).

Knowing how many diseases occur and which are the most frequent necessarily requires standard classification and measurement criteria. This is true regardless of whether we adopt a realistic or empiricist perspective from the philosophical point of view since in both cases the question is how to recognize diseases that affect patients and how to distinguish between them.

This would seem to be easy since it would involve verifying the correspondence between the data of the affected patient obtained from their medical history and an exploration and those that characterize a certain disease. This would be obvious if such patient data were pathognomonic, that is if all patients who displayed them suffered the same affliction.

Naturally, this requires the existence of a catalogue or repertoire of diseases, a classification that includes at least those universally accepted until now—because some reappear, they are called emerging diseases—and above all one that indicates, if possible in a mutually exclusive way, that a certain clinical case with its properly observed signs and symptoms is or is not a certain disease.

Unfortunately, the classifications of diseases in use not only do not include the necessary taxonomic requirements, but the different chapters they are organized into do not even meet the mutually exclusive condition that would be required of any biological taxonomy. The basic format of the current international classification has chapters that include diseases according to their etiology, such as infectious diseases, together with others that group them according to their manifestations such as cancer and, of course, those that correspond to anatomical location: respiratory, circulatory, renal diseases, etc.

It is worth remembering that we are talking about diseases, that is, nosological entities, rather than disease as a concept. And to specify the nature of disease as a concept is no easy task. In fact, other names such as syndrome are often used, which would be more ambiguous, although in some cases it is as or more practical than a disease that is indisputably considered as such. An example of this is acquired immunodeficiency syndrome, which, despite the great variability of its manifestations, is a consequence of a susceptible host being infected by a certain virus.

Just a few decades ago, disease was not mentioned unless there were some negative alteration in the capacity for normal functioning, pain, functional impotence, polyuria, haemorrhaging, etc. Today, on the other hand, many people recognize themselves as being sick and affected by more than one chronic disease without the perception of their own health being bad. In some cases this is because alterations are detected when there are still no clinical manifestations, and in others it is because risk factors that increase the probability of presenting a disease are confused with already being ill. This is the case with many patients who have high blood pressure without their organs being affected or those affected by high cholesterol and even people with osteoporosis, to say nothing of those who simply have osteopenia.

In this respect, it may be useful to increase the specific meaning of words often used as synonyms. For example, when we talk about pathological entities (diseases), illness or sickness. According to Susser (1990), disease (the pathological entity) would be the more academic and objective term, while illness would refer to the patient's subjective perception of the experience and sickness would refer to the real social recognition of a disease whose identification and diagnosis correspond to medicine.

The above reveals the insufficiency of language in reflecting and understanding reality. Even if we admit, first with Locke and then later with Popper, that nature exists independently of us, we are the ones who perceive it, with the limitations of our senses, our knowledge and the intelligence with which we produce it. On this basis, it is logical to aspire to explanations that transform reality, such as those that are the result of identifying a cause of disease or good health and are capable of modifying it effectively and beneficially even if we do not fully understand what is happening.

The natural history of disease (Centers for Disease Control and Prevention 1992) could provide us with some clues to distinguish between situations that will eventually have negative consequences from those that will not. But the fact is that in many cases the evolution of some objective disorders such as dysplasia in the case of cervical cancer is not necessarily progressive. Of course, the same concept of natural history already assumes an acceptance that diseases have their own existence, regardless of the circumstances and the patients who suffer them, which does not seem very common if we discount certain cases such as measles. The vast majority of humans vulnerable to infection by the measles virus develop perfectly recognizable symptoms. However, this is not the case with other equally or more frightening viral infections, such as polio, for example. Many people are exposed to the polio virus and have no reaction whatsoever, and among those who do develop symptoms, most fortunately only suffer minor problems. Only a small minority suffer an affectation of the motor neurons, which in some cases can lead to death and in others to disabling after-effects.

Even if diseases do tend to display some regular features in their manifestations, a propensity to progression—natural history—and a physiopathological alteration that explains observable changes—probably the most recognized feature from a biological perspective, we do not have a general theory of disease that guides us. And neither would it seem easy to design one. When it first emerged, the so-called germ theory proposed by Henle and Koch raised hopes among positivists and mechanists of a potentially satisfactory explanatory model. However, in practice it continues to lack any real credibility, as illustrated by the controversy generated around the viral etiology of AIDS and recourse to Koch's famous postulates as evidence (Duesberg 1988). The limitations of this unicausal and linearly deterministic model are notorious, particularly when the nosological entities we are faced with are the so-called chronic diseases, including mental illnesses.

Attempts to construct a general explanatory causal model based on epidemiology have also failed to achieve practical applications beyond broadening the focus from the so-called black box epidemiology, which in fact means recognizing our ignorance, to the so-called Chinese box epidemiology, latterly known as eco-epidemiology (Susser and Susser 1996; Susser 2004). It would therefore seem more practical to adopt an eclectic attitude and look to improve health and control diseases, even if we are not sure we have understood the question definitively.

Either way, classifying certain situations as diseases is no mere semantic issue or a speculative aside unconnected to the events of everyday life. Not by any means. Considering a certain situation a disease has practical consequences. Some in the domain of more personal relationships; but others in the domain of society and politics. Although it is not the object of these considerations to recognize the objective existence of diseases, it does mean accepting that suffering from them goes beyond the will or behavior of those affected, even if the will and attitudes of patients and the society they live in do modulate the experience and make it more or less bearable. We are not talking about simulation or fraudulent behavior, however. The so-called attention deficit hyperactivity disorder would be an illustrative example, and fibromyalgia or chronic fatigue even more so. On another level we find so-called risk factors, including osteopenia—a phase prior to osteoporosis—which when treated as disease lead to inadequate medicalization and consequently iatrogenesis. Medicalization is also debatable in such cases as so-called erectile dysfunction and many other situations that require medical interventions without there truly being a pathology, such as androgenic baldness, freckles or ugliness. Of course, there are even more complex situations such as transsexuality that, while claimed to be non-pathological, requires the adoption of medical and surgical interventions that are even publicly funded.

Health: Beyond the Absence of Disease

Beyond the common definition that we find in general dictionaries and that, in general, accentuates the absence of disease as an essential attribute, even if in many cases it is also complemented by a mention of the proper functioning of the organism, which involves acceptance of a certain margin of fluctuation around the physiological average, there are definitions of health other than the absence of disease. Among the best known is that of the WHO, officially adopted by the first world health assembly in 1948 and proposed by the Croatian health specialist Andrija Stampar (Tejada de Rivero 2003): "health is a state of complete physical, mental and social well-being and not merely the absence of diseases or infirmity". It is a much debated definition, although in its favour it can be argued that it has led to significant changes in the field of healthcare and health policies (Callahan 1973). Firstly, because health can be understood as an attribute, a characteristic, a state or a positive experience that is not generated simply through the prevention or cure of disease; that is, it requires some action to promote it, rather than mere preventive or restorative actions. And also because it makes the three inseparable dimensions of the biological nature of human beings explicit: the physical or somatic, the psychological, mental or spiritual and the collective or social, all of which are capable of experiencing different degrees of well-being and therefore of being qualified as healthy or not.

Unfortunately, well-being is a very subjective characteristic that can be enjoyed in very diverse, even alienating, circumstances since the feeling of well-being is compatible with ignorance, intoxication and even idiocy. It would therefore be useful to identify those characteristics of health that allow an operational approach, both from the individual and the collective perspective, the latter being the one that interests us here since our aim is to meet the health needs of populations.

We may then consider health as the ability to function better, faster, higher, and stronger, as in the Olympic motto "*citius, altius, fortius*". As we know all too well, however, the effort athletes make does not really lead to improved health, and in fact professional sport implies a high risk to competitors' health. As Kingma points out (Kingma 2012), beyond what an organism is capable of doing here and now, it is more logical to consider what it is capable of doing throughout its life, meaning that health could be the ability to acceptably survive in the environments we live in and overcome the challenges we face, an approach that recalls the definition posited by René Dubos, for whom health is "the expression of the extent to which the individual and the social body maintain in readiness the resources required to meet the exigencies of the future" (Dubos 1959).

Adaptive capacity can also be expressed as resilience, a characteristic that allows objects to recover their position after having been displaced, like a tilting doll that has a counterweight at the base and always returns to an upright position, or reeds, which bend but do not break. In addition to being an individual aptitude, resilience is also applicable to ecological systems (Mumby et al. 2014) and human communities (Kirmayer et al. 2009). As a concept it is related to social capital, which, in a certain way, constitutes one of the community assets of health (Ledogar and Fleming 2008), or one of the collective determining factors of a population's health, which also traditionally includes the environment. And if health depends in part on the environment, both physical and social, one way to improve or promote it will be through healthy environmental and social policies.

This salutogenic conception, in contrast to the pathogenic perspective characteristic of medicine, we owe to Antonovsky. Recognizing the factors associated with survival in the Nazi concentration camps (Antonovsky et al. 1971), he developed a model to apply to the promotion of community health (Antonovsky 1996; Lindstrom and Eriksson 2006).

Health and disease refer to closely related aspects, which only come into conflict if we opt for a negative conception of health. And, as the definition of disease has been broadened to a great extent, they would not altogether be antonyms, at least in some cases.

Implications for Formulating Priorities and Designing Policies

As stated at the beginning, the establishment of priorities is the most important factor, from an ideological perspective, in any political program, since it means ordering the problems to be addressed or needs to be met according to their level of priority.

Therefore, it is necessary to define such needs in an operational way, which can be done by using different criteria, the most common of which are health problems expressed as diseases. These can be weighted according to their frequency and severity, their impact in terms of health deficit, of direct and indirect economic costs, of alarming the population, etc. Epidemiology provides us with indicators to measure the frequency of diseases (prevalence and incidence rates) and their severity (lethality and mortality rates). However, in order to compare the consequences of some diseases in relation to others, we can use the concept burden of disease (GBD 2016 Causes of Death Collaborators 2017), which comprises a set of morbidity and mortality estimates in populations that allows us to quantify loss of health status due to different pathologies, injuries and risk factors in a comparative way by variables related to individuals, time and place. The unit of measurement for burden of disease is disability-adjusted life year, or DALY.[1]

But other criteria of need or, better still, of healthcare or health aims may also be considered, such as producing positive health. In this regard, in addition to recognizing which aspects of the health of people and populations we would like to improve, we must also know how such objectives can be achieved.

The pathogenic vision is the one traditionally adopted by medicine. The idea of the natural history of disease illustrates this clearly. It is a perspective whereby even risk factors are treated as pre-diseases, when their genuine meaning is different, since what they imply is a higher probability of presenting the disease, rather than an indication of its certain appearance, while not being exposed to the risk factor does not imply any guarantee of not suffering it in the future. Therefore, recognizing a risk factor should be seen as an opportunity for prevention and not the perception that we are already sick and require treatment. The pathogenic perspective, on the other hand, is one that interprets the promotion of health basically as its maintenance and the prevention of disease. It is a vision with a long history behind it and many relevant contributions. Not to mention the spectacular nature of some of its achievements. But it also has its limitations. In one sense because the history of medicine has

---

[1]The disability-adjusted life year (or DALY) is an indicator of the burden of disease. It was originally designed by Harvard University for the World Bank and adopted by the WHO in 2000. It takes into account an older indicator, potential years of life lost (due to premature death) but only those that would be lived in good health, so that mortality and morbidity are measured at the same time by means of a single and comparable indicator. A DALY is equal to a lost year of healthy life. Suitable for measuring the impact of chronic diseases, it weighs the importance of low mortality diseases such as mental and neurological ones.

particularly represented the history of disease, even when we do not know for certain what it is. And this does not precisely enable us to improve or increase health. In another sense, perhaps more concerning, because exaggerated, superfluous or futile medicalization is also dangerous and contributes to producing adverse effects on the health of individuals and communities. And such damages are not attributable solely to errors or negligence.

The salutogenic vision, on the other hand, considers health to be something more than and sometimes even different from the mere absence of disease. To the point that while suffering from one or more chronic diseases the perception of one's health can be good or even very good. This is because what it contributes is the ability to resist difficulties, stress and anxiety, including diseases, and to function better and adapt to the environment to a greater or lesser degree.

From the point of view of policies and prioritization, adopting the salutogenic vision is a decisive step. It is therefore a question of identifying those factors, basically in the physical and social environment (education, for example) that are associated with greater resilience and considered positive collective determinants of health so that they can be formulated as policy objectives.

However, in the development of procedures and technical criteria for establishing priorities in the health sphere, much more importance is awarded to more traditional criteria for defining health needs (Tromp and Baltusen 2012), in particular morbidity and mortality, sometimes expressed in DALYs. That said, it is true that there has been a tendency of late to value the efficiency and equity of potential interventions, most of which are carried out within the strict scope of the healthcare system (Guindo et al. 2012). And this favours the corporate interests of healthcare institutions, both public and private, when it comes to developing effective health promotion policies.

For a Healthy Overall Health Policy: The Role of Public Health as a Social Institution

In 1986, the Ottawa charter called for a reorientation of the healthcare system and the promotion of healthy public policies. Twenty years on, when presiding over the European Union, the Finnish government led the "Health in All Policies" initiative (Ståhl et al. 2006), perhaps thanks to the prestige it had acquired through North Karelia's community-wide heart attack prevention project (Puska 2002). This approach, in theory at least, opened the door to establishing health priorities beyond those that correspond to health departments, which are more vulnerable to corporate interests. It was an intersectoral strategy (Wismar et al. 2012) to be used to design a set of healthy public policies that actively involve the sectors of society and public administrations that influence the health of populations, such as road safety, safety in the workplace, food safety, etc. or in a more global way, the environment, which as well as sustainable must be as healthy as possible for people (Kickbusch 2013).

Although this approach to health in all policies has been adopted with greater or lesser vigour in some countries, it is worth highlighting experiences in local and municipal spheres (World Health Organization 2013; Gase et al. 2013), since they tend to involve healthcare mechanisms related to public health. These can play a coordinating role with sectors of society that effectively contribute to protecting and improving collective health and can therefore also be considered part of public health. In fact, Winslow's 1920 definition of public health refers to the organized effort of society—not only the healthcare system—to preserve and promote collective health (Winslow 1920).

This link between public health in the healthcare system and the public health of society as a whole could facilitate the development of a cooperative, intersectoral and participatory strategy between health and society, whereby the priorities of public policies are oriented not only towards the prevention of disease and care for the sick but also the improvement of positive social determinants of health.

References

Antonovsky, A. (1996). The salutogenic model as a theory to guide health promotion. *Health Promotion International, 11*, 11–18.

Antonovsky, A., Maoz, B., Dowty, N., & Wijsenbeek, H. (1971). Twenty-five years later: A limited study of the sequelae of the concentration camp experience. *Social Psychiatry and Psychiatric Epidemiology, 6*, 186–193.

Benoit, F. (2013). *Public policy models and their usefulness in public health: The stages model.* Montréal: National Collaborating Centre for Healthy Public Policy. Accessed from http://www.ncchpp.ca/165/publications.ccnpps?id_article=966

Callahan, D. (1973). A WHO definition of health. *The Hasting Center Studies 1*(3), 77–87. Accessed from http://www.jstor.org/stable/3527467

Centers for Disease Control and Prevention. (1992). *Principles of epidemiology* (2nd ed.). Atlanta: U.S. Department of Health and Human Services. Accessed from https://www.cdc.gov/ophss/csels/dsepd/ss1978/lesson1/section9.html

Dubos, R. (1959). *Mirage of health.* New York: Harper & Row.

Duesberg, P. (1988). HIV is not a cause of AIDS. *Science, 241*, 514–517.

Gase, L. N., Pennotti, R., & Smith, K. D. (2013). Health in all policies: Taking stock of emerging practices to incorporate health in decision making in the United States. *Journal of Public Health Management and Practice, 19*, 529–540.

GBD 2016 Causes of Death Collaborators. (2017). Global, regional, and national age-sex specific mortality for 264 causes of death, 1980–2016: A systematic analysis for the Global Burden of Disease Study 2016. *The Lancet, 390*, 1151–1210.

Goetghebeur, M. M., Wagner, M., Khoury, H., Levitt, R. J., Erickson, L. J., & Rindress, D. (2008). Evidence and value: Impact on decision making – the EVIDEM framework and potential applications. *BMC Health Services Research, 8*, 270.

Guindo, L. A., Wagner, M., Baltussen, R., Rindress, D., van Til, J., Kind, P., & Goetghebeur, M. M. (2012). From efficacy to equity: Literature review of decision criteria for resource allocation and health care decision making. *Cost Effectiveness and Resource Allocation, 10*, 9.

Hanlon, J. J. (1974). *Public health: Administration and practice.* Saint Louis: C.V. Mosbi Co.

Heath, I. (2007). In defence of a national sickness service. *British Medical Journal, 334*, 19.

Kickbusch, I. (2013). Health in all policies. *British Medical Journal, 347*, f4283.

Kingma, E. M. (2012). *Health and health promotion*. Eindhoven: Tecnische Universiteit Eindhoven. Accessed from https://pure.tue.nl/ws/files/3462442/kingma2012.pdf

Kirmayer, L. J., Sehdey, M., Whitley, R., Dandenau, S. F., & Isaac, C. (2009). Community resilience: Models, metaphors and measures. *Journal of Aboriginal Health*, 62–117.

Ledogar, R. J., & Fleming, J. (2008). Social capital and resilience: A review of concepts and selected literature relevant to aboriginal youth resilience research. *Pimatisiwin, 6*(2), 25–46.

Leeflang, M. M., Deeks, J. J., & Gatsonis, C. (2008). Bossuyt cochrane diagnostic test accuracy working group. Systematic reviews of diagnostic test accuracy. *Annals of Internal Medicine, 149*(12), 889–897.

Leppo, K., & Ollila, E. (2013). *Health in all policies: Seizing opportunities, implementing policies*. Helsinki: Ministry of Social Affairs and Health, Finland.

Lindstrom, B., & Eriksson, M. (2006). Contextualizing salutogenesis and Antonovsky in public health development. *Health Promotion International, 21*, 238–244.

Makary, M. A., & Daniel, M. (2016). Medical error—The third leading cause of death in the US. *British Medical Journal, 353*, 2139.

Marmot, M. (2005). Social determinants of health inequalities. *Lancet, 365*, 1099–1104.

Menten, J., & Lesaffre, E. (2015). A general framework for comparative Bayesian meta-analysis of diagnostic studies. *BMC Medical Research Methodology, 15*, 70. https://doi.org/10.1186/s12874-015-0061-7.

Mumby, P. J., Chollett, I., Bozec, Y. M., & Wolff, N. H. (2014). Ecological resilience, robustness and vulnerability: How do these concepts benefit ecosystem management? *Current Opinion in Environmental Sustainability, 7*, 22–27.

Naaktgeboren, C. A., Ochodo, E. A., Van Enst, W. A., de Groot, J. A. H., Hooft, L., Leeflang, M. M. G., Bossuyt, P. M., Moons, K. G. M., & Reitsma, J. B. (2016). Assessing variability in results in systematic reviews of diagnostic studies. *BMC Medical Research Methodology, 16*, 6. https://doi.org/10.1186/s12874-016-0108-4.

Pineault, R., & Daveluy, C. (1986). *La planification de la santé: Concepts, méthodes, stratégies*. Montréal: s.n. (éditeur).

Puska, P. (2002). Successful prevention of non-communicable diseases: 25 year experiences with North Karelia Project in Finland. *Public Health Medicine, 4*(1), 5–7.

Segura, B. A. (2014). Recortes, austeridad y salud. Informe SESPAS. *Gac Sanit, 1*, 7–11.

SESPAS/OMC Iatrogenesis. (n.d.). Analysis, control and prevention. Position paper. Accessed from sespas.es/.../English-IATROGENIA-DOCUMENTO-SESPAS-OM...

Ståhl, T., Wismar, M., Ollila, E., Lahtinen, E., & Leppo, K. (Eds.). (2006). *Health in all policies. Prospects and potentials*. Helsinki: Ministry of Social Affairs and Health and European Observatory on Health Systems and Policies.

Susser, M. (1990). Disease, illness, sickness; impairment, disability and handicap. *Psychological Medicine, 20*, 47.

Susser, E. (2004). Eco-epidemiology: Thinking outside the black box. *Epidemiology, 15*, 519–520.

Susser, M., & Susser, E. (1996). Choosing a future for epidemiology: II. From black box to Chinese boxes and eco-epidemiology. *American Journal of Public Health, 86*, 674–677.

Tejada de Rivero, D. A. (2003). Alma-Ata revisited. *Perspectives in Health, 8*, 3–7. Accessed from http://www.croatia.org/crown/articles/9595/1/Andrija-Stampar-is-the-father-of-the-World-Health-Organization.html

Tromp, N., & Baltusen, R. (2012). Mapping of multiple criteria for priority setting of health interventions: An aid for decision makers. *BMC Health Services Research, 12*, 454.

Winslow, C. E. A. (1920). The untilled fields of public health. *Science, 51*, 23–33.

Wismar, M., McQueen, D., Lin, V., Jones, C. M., & Davies, M. (2012). Intersectoral governance for health in all policies. *Eurohealth Incorporating Euro Observer, 18*(4), 3–7.

World Health Organization. (2013). *Demonstrating a health in all policies analytic framework for learning from experiences: Based on literature reviews from Africa, South-East Asia and the Western Pacific*. Geneva: World Health Organization. Accessed from www.who.int

Critical Epidemiology in Latin America: Roots, Philosophical and Methodological Ruptures

Jaime Breilh

Abstract Epidemiology as any other scientific work dealing with social and human wellbeing runs under great social pressure. Its conceptual and practical tools have developed in concrete historical conditions facing evident social inequity. In recent times, neoliberal economic acceleration, through the ongoing fourth industrial revolution, has globalized massive unhealthy processes and expanded economic and political influences.

In Latin America critical epidemiology evolved from the early 1970s social medicine movement. However limited by its subordinate position with respect to mainstream medicine, this insurgent academic position grew as a counteractive intellectual and political tradition that based its initial impulse on epistemological clarity and activism. Contrary to the linear functionalist thinking of conventional medical science—that separates the understanding of health conditions from their social and cultural contexts—critical social medicine broke away from that positivist framework.

This chapter discusses the historical roots, epistemological and methodological ruptures of critical epidemiology. It describes the formation of the social determination of health paradigm of dialectical, complex, critical thinking with its potential for an emancipatory perspective of health reform. In the same breath, it presents a sound critique of the "knowledge illusion" of linear, reductionist causal thinking, while assuming transformative science as the reference for an integral scientific objectiveness that confronts and redefines the traditional notion of objectivity and recovers the active transforming role of the subject in science. It proposes complex thinking to decipher the real requisites of rigorous epidemiology, which does not only depend on the accuracy, reliability and validity of its empirical methods,

I should like to express my grateful thanks to Michael Harvey (Professor at the University of California) and Maria Cristina Breilh for their rigorous and generous support in the translation of some of my original texts in Spanish the conceptual explanation and synthesis of recent works being prepared in English. Their contribution was decisive for the preparation of this chapter's original English version.

J. Breilh (✉)
Universidad Andina Simón Bolívar, Quito, Ecuador

J. Vallverdú et al. (eds.), *Philosophical and Methodological Debates in Public Health*, https://doi.org/10.1007/978-3-030-28626-2_3

calculations and observations but on a rigorous and multidimensional/complex/transdisciplinary/intercultural understanding of society and health.

Keywords Epidemiology · Critical science · Complex thinking · (Health) epistemology · Methodology

The Challenging Context of Epidemiological Science

In the Latin American South, since the early twentieth century, the visible signs of extreme political authoritarianism and social inequity, as well as the growing unfairness of the broader world economy, inspired a culture of critique and resilience within the region that left a deep impression upon its health scientists. One institutional response of academics to this widespread inequality appeared in the mid-1970s as the social medicine movement. A counteractive intellectual and political tradition that based its strength on epistemological clarity and activism, social medicine was however limited by its subordinate position with respect to mainstream, dominant, and much more generously financed approaches to health science. Despite this fact, both in public and small private research settings, new ideas and agency flourished from this movement.

In the Global North, the historical and vital counterhegemonic traditions of critical public health and social medicine—comparatively stronger in their technical foundations and institutional resources than their southern counterparts—were nonetheless subsumed by the dominant positivist and functionalist public health paradigm. Globally, resources and funding for health research have been mainly concentrated in the basic sciences and in applied clinical and surgical domains, whose findings are more readily commoditized. Meanwhile, public health research faces constraints that hinder the recognition of its theoretical and methodological sophistication and of its practical implications for transformative health prevention and promotion. The driving force of this commoditized science and functionalist research paradigm is the global expansion of transnational corporate control over health care, research, and teaching organizations.

The absurdity of this system becomes more evident when recent rigorous reports condemn supposedly objective mainstream medical research as an overvalued waste. A well-documented denouncement of this "biomedical bubble" has characterized its biased priorities, lack of diversity and systematic waste of financial resources. It has explained this bubble as a speculative fraud that overestimates the effect of certain drugs and rules out investment and academic concern on the real health problems of society (Jones and Wilsdon 2018).

But this corporate influence also puts pressure on public health entities, their scope of concern, and their mandate for social change. As Waitzkin notes, "An increasingly transnational corporate health care industry... aggressively aims to exploit the gaps left open by underfunded or nonexistent public provision, furthering commodification and fragmentation..." (Waitzkin and Working Group for Health Beyond Capitalism 2018, p. 239).

It is within this contradictory and contested position in society that epidemiology, public health's so-called "diagnostic arm", must operate. Under great social pressure, epidemiology is called on to produce objective assessments of social wellbeing and of the role of social factors in shaping population health. Epidemiological statements and indicators are also considered barometers of the health status of the population degree of healthiness and wellbeing obtained by social policies. These statements explicitly and implicitly evaluate the effectiveness of public policy and governmental strategies, and thereby justify or cast doubt on those individuals and parties in power. Hence, epidemiological reports constitute an outstanding example of politically sensitive data systematization, frequently submitted to methodological manipulations to produce functional information by separating facts from their contextual determinations.

It is important to emphasize the fact that conventional epidemiological paradigms used within public administration continually divorce the development of health indicators from their social and cultural contexts. Epidemiologic reason operates under the premise that health standards alone suffice as evaluation tools to assess the success of social policies. Some classical indicators of changes in morbidity and mortality rates or the degree of health services coverage are considered, in themselves, as the gold standards for weighing the effect of public policy and governance. Therefore, such evaluations are often limited to the analysis of isolated programs and services provision and to classical epidemiological indicators. From this perspective, when a society goes from "bad rates" to "less bad rates", success is declared.

In Latin America, the myth of a "good extractivism" that yields funds for health investment also obscures the contemporary understanding of and chances for healthy living. The practice of extractivism has historically appeared in countries with an abundance of valuable natural resources. Extractivism implies a form of governance that subordinates the logic of the State to the strategic interests of powerful international corporations involved in natural resource extraction. It entails aggressive investment in the extraction of exportable, non-renewable goods. Through oil and mining concessions and agribusiness, companies negotiate forms of accelerated capital accumulation that operate through convergence of profit-based applications of technology (Ribeiro 2016), dispossession of life goods (i.e. land, water, genome, and in general of the biota) (Harvey 2003), and frequently the opportunistic exploitation of disasters (Klein 2008). Under extractivism, countries become trapped in a perverse logic through these three exploitation pathways, which paradoxically converts their abundance into impoverishment (Breilh and Tillería 2008). In recent times extraction also operates in the virtual world through the oligopoly of virtual data extraction platforms that convert mega personal databases into lucrative merchandise (Subirats 2019).

In a manner that reminds us of the recent debates on political fake truth, the discourse of "good" extractivism that pays for social expenditures has become common over the past two decades, especially among self-proclaimed "progressive" Latin American governments. The notion of good extractivism implies a form of governance that finds its legitimacy by distancing itself in the public's memory from the open privatization that has characterized the neoliberal period of the late twentieth and early twenty-first century. Furthermore, these "progressive" governments pursue a strategy of aggressive public-private infrastructural development and administrative modernization while retaining targeted, means-tested service delivery policies for the extremely poor. This mainstream epidemiology concerns itself with the visible peak of the metaphorical iceberg, while strategically eliding the deeper pathogenic processes of extractivism, capital accumulation, and ecological damage that are at work.

Therefore, the fourth industrial revolution of capitalism implies an explosive convergence of new productive technologies to accelerate capital accumulation and to reduce production costs. An array of applications in robotics, nanotechnology, biotechnology, big data operations, hypermedia and artificial intelligence constitutes a powerful and uncontrolled industrial arsenal. Healthwise, this avalanche of greedy applications has been described as a technological tsunami. It holds unpredictable risks for human and ecosystems health. It expresses a philosophy of unrestrained energy expenditure (i.e. one global instantaneous ultra complex bitcoin operation consumes as much energy as a normal average household would in 1 week). In this avalanche, synthetic biology implies unhindered genetic sequencing; the boom of big data management is not only used to penetrate intimate personal information for marketing purposes but also for personal political control. As a whole, this avalanche signifies an unbounded diversification of potential and real impacts on the present and future of social wellbeing and sovereignty (Ribeiro 2016).

The twenty-first-century paradox is that this big business control of new technologies coexists with overt, rapacious, pre-capitalist labor exploitation strategies applied in subordinated regions and countries of the South.

Either way, the urge to raise profit implies the expansion of clearly unhealthy labor and domestic environments. Capital accumulation demands highly specialized and continuous production processes on a big scale. In the case of agricultural extractivism, it involves the control over land, water, seeds, and genetic resources. Land monopoly also signifies territorial monotony, because profitable low-cost production processes only operate efficiently in immense monotonous one-crop landscapes. Biodiverse, multiple crop territories, are economically inefficient when analyzed through this shortsighted perspective. Under those conditions, it has been said that "biodiversity amounts to bad corporate business" (Bartra 2009).

From this perspective, it becomes clear that epidemiology now operates in social contexts characterized not only by the evident social inequity that has accompanied the market society for centuries, but also by new health hazards and deteriorating ecosystems. The economic interests of corporations clash with the strategic interests of populations demanding good living conditions. This struggle is shaped by social interests who are defined by their type of relation with exploitation and capital accumulation, that is, by their class position. In this regard, this scientific domain shares with all others an inherent symbolic aspect, and is thus "... a transformed, subordinated, transmuted, and sometimes unrecognizable expression of the power relations of a society" (Bourdieu 1998). In our case, those relations entail the imposition of a system of social exploitation and of the exploitation of nature, a system which reproduces unsustainable, inequitable, and unhealthy societies. In what follows, I describe an innovative, intercultural and transdisciplinary epidemiological science that is defined by a number of ethical, theoretical, epistemological and methodological breaks with hegemonic mainstream epidemiology. I refer to the conceptual core of this innovative science as *social determination of health.*

The Roots and Ethics of a Critical Scientific Tradition

The *social determination of health* is one of three central categories of our proposal for a critical epidemiology, described extensively for the first time in 1977 (Breilh 1977) and subsequently in several works by other authors of the Latin American social medicine and collective health movement. Together with the categories of *social reproduction* and *society-nature metabolism*, the social determination of health has formed, since the beginning of our work in the 1970s, a theoretical base with which we have built a process of rupture with the old empirical-functionalist public health paradigm. The social determination of health paradigm incorporates new contents and practical articulations. It alludes to a different epistemological, conceptual and methodological-instrumental development of epidemiology.

Latin American critical thought drew from two important sources of critical thinking within epidemiology: (1) the revolutionary ideas about determination of health from historical actors who inspired the anti-colonial struggle of eighteenth-century Latin America, such as Eugenio Espejo in Ecuador (Espejo 1788–1994–) (Espejo 1994); and (2) the revolutionary works of nineteenth-century Europeans, such as Frederick Engels and Rudolf Virchow.

Virchow's potent call for action in his groundbreaking nineteenth-century report (Virchow 1848) clearly stated that "preserving health and preventing disease requires 'full and unlimited democracy' and radical measures rather than 'mere palliatives'".

These works, as Howard Waitzkin argues in his magnificent critique of medicine and public health in *Medicine and Public Health at the End of Empire*, created the perspective of social medicine; and documented the impact of early capitalism, basically before the empire's expansion (Waitzkin 2011, pp. 9–10).

In that same book, Waitzkin highlights the pioneering work of critical thought from the Global South by Salvador Allende (1939). Allende's report, *On the Chilean Socio-medical Reality*, recognized the relationship between political economy, disease, and suffering by focusing its "causal" gaze on the role of empire, underdevelopment, and the need for structural change and for change in the life of the proletarian classes as the fundamental solution to health inequality (Waitzkin 2011). That is to say, this second source of critical epidemiology did not come solely from the works of nineteenth-century Europeans, but it had other pivotal proponents in Latin America whose contributions, often silenced by official history, must be rescued.

As a result of the turbulence and social awareness of the early decades of the twentieth century, there was a consolidation of revolutionary social thinking that penetrated thought about health and health inequalities. This consolidation favored the emergence of other figures dedicated to critical thinking in epidemiology, such as Ricardo Paredes (1938), who as a physician rigorously studied the social, workplace, and health conditions of workers of a mining company. Paredes later published a remarkable and pioneering epidemiological essay on the determination of health in imperial mining which, supported by robust sociological thought and statistical evidence, provided a rich analysis of the destruction of health and the environment in Ecuador (Paredes 1938). The works of Ramón Carrillo (1951) are also fundamental to the consolidation of this perspective. These include the Synthetic Public Health Plan for Argentina, in which Carrillo situates epidemiological thought as central to the search for equity and to creating a profound vision of disease prevention. In North America, the guiding thread of critical epidemiological reasoning of the social and integral system was expressed in works such as those of Henry Sigerist. *Civilization and Disease* made a forthright pioneering contribution to the broadening of health science by incorporating the role of economics, culture, philosophy, the arts and an interdisciplinary approach in the understanding of health (Sigerist 1945).

George Rosen's *History of Public Health* made crucial contributions to the progressive understanding of the origins, historical transformations, and socially determined conditions of public health (1958) (Rosen 1958). His works inspired many workshops on the critique of functionalist Public Health.

Thus, it was both in Latin America and the Global North where the conditions that gave birth to the most advanced modern critical epidemiology were incubated. And under the social influence of the 1970s, the conditions were created for several nuclei of social medicine in Latin America to focus on the relationship between the capitalist social order and health, a scenario in which I was able to propose and systematize the use of the concept of '*social determination* of health' in epidemiology. This concept was fully developed in my thesis that proposes and develops in a direct, detailed, and systematized way, a critique of the empirical-functionalist paradigm of epidemiology, and proposes the concept of *social determination* of health as a tool to explore the relationship among social reproduction, modes of living, of sickness and of death (Breilh 1977).

It should be noted that Latin American critical epidemiology, albeit forged in academic settings, has always had as its source of inspiration the struggle of our peoples. The purpose of Latin American critical epidemiology is neither to decipher empirical connections to promote reformist measures and represent the entelechy of health in a structurally pathogenic society, nor do we aim to replace the logic of explanation with the mere prediction of empirical phenomena. Critical epidemiological science cannot be anything but radically emancipatory. Finally, the notion of social determination of health, like any idea of rupture, is not the isolated product of one person's musings, but rather the expression of critical thought that had seized the Latin American social medicine movement (now better known as collective health); its origins lie in the convergence of collective concerns of mobilized groups, the development of new tools of social analysis, and a project expressly concerned with the transformation of the obsolete paradigm of empirical-functionalist public health of the 1970s.

Interpretative and Philosophical Challenges

The social determination of health paradigm is a transdisciplinary intercultural framework.

It has evolved from two main traditions of thought: critical social scientific thought developed in academic settings and counterhegemonic thinking of Latin American peoples' movements revealed through manifestos and position papers. In regard to the former, it is most closely associated with the materialist critical realist tradition of Marx and other Marxist thinkers, although it draws broadly from authors of political economy, the philosophy of praxis, critical epistemology, and critical ecology, to name just a few.

Along these lines, the social determination interpretative model for critical epidemiology is informed and inspired by convergent critical epistemologies and academic traditions: Marxist political economy and critical realism that claim for a profound change of social relations which reproduce social subordination and inequity and impede a fair and democratic distribution of wealth and opportunities for wellness (i.e. to make real "from each according to his ability, to each according to his needs"); Foucault's constructivist demands to unveil mechanisms of cultural coercion, to question truth as it operates through power, and to reinvent the scientific certainties that hegemonic science has imposed (Foucault et al. 2007); the Frankfurt Schools' critical reflections on reason; the powerful anticolonial epistemologies of decolonial thinkers; the groundbreaking contributions of indigenous peoples and of the feminist and gender struggles. Other authors like Weber and Bunge provide important methodological guidelines for the critique of positivist causal methodology. And thinkers like Schmidt, Enzensberger, Marcuse, and Lowy have offered substantial clues for the reconceptualization of nature and ecological relations that shape the eco-social determination of health. Concerning the reconceptualization of social and ecological relations, the crucial participation of non-academic knowledge refers to the groundbreaking, inspiring, and emancipatory construction of values and ideas about intercultural solidarity, healthiness, and modes and relations of good living emerged from the struggles of the indigenous peoples of Ecuador.

As a whole, the key issue of transformative critical science is to connect with the peoples' problems and strategic needs. This does not mean only taking clear stands on practical political issues. It also and fundamentally means moving away from official, functional, bureaucratic science.

To do so in epidemiology, we must liberate academic work and research from the dominant logic of causalism with its linear reductionist mono-cultural methodology and build on the social determination of health paradigm. This epistemological operation parts from a productive and innovative synthesis of the aforementioned currents of critical thought. This new approach is essentially intercultural and ultimately seeks to explain the relationship between capitalist society, the modern civilization that supports it, and collective health. These innovations can be thought of as a series of breaks with traditional or hegemonic epidemiology, which also indicate the need to understand the theoretical and political misguiding and incomplete arguments of the social "determinants" of health paradigm as defined by its founders (Marmot and Wilkinson 2006).

Within well-informed progressive academic scenarios of the South, this supposedly "new" model appeared as an essentially contradictory perspective. It came into view three decades after the initial Latin American works on social determination were published in two or three languages. This Latin critical epidemiology bibliography was clearly familiar to progressive scholars of the North which published important reviews about our work in high impact peer-reviewed English journals. But they were bluntly ignored in Eurocentric scenarios. Historical epistemology has been dealing with this to understand this academic blindness of certain conventional epidemiologists of the North and has interpreted the absence of Latin American bibliography in the social determinants exegesis, not as a result of mere language barrier but as an emblematic case of biased Eurocentric knowledge. Willingly or not, this omission produced an epistemicide or banishment of sound anti-establishment science coming from the South.

Nevertheless, what matters for the purpose of this paper is that, after three important international scientific meetings dedicated to the debate between the social determinants and the social determination approaches—organized in Brazil (2010), in México (2011) and in Colombia (2015)—it became clear that the former operates in the functional logic of causalism (i.e. "causes of the causes"); it pertains to a reformist institutional perspective; and it cherishes redistributive governance as its leitmotiv. Whereas, the social determination of health paradigm performs in the framework of dialectical, complex, critical thinking; assumes an emancipatory perspective of health reform; entails a critique of market economy and of individualistic consumerist civilization; and aims at the transformation of dominant social power relations that reproduce unhealthy collective conditions.

This debate has enriched the critique of risk factor epidemiology amply criticized in Latin American bibliography because of its total reliance on probabilistic reductionism (Ayres 1997; Almeida Filho 2000; Breilh 2003).

In tone with Virchow's demand for radical measures as opposed to palliatives, critical epidemiology alludes to diverse, multidimensional, contradictory driving forces that produce healthy or unhealthy modes of living, according to historical social relations; each group or person being affected according to its social position in the hierarchic structure of the capitalist society. Social power relations determine the modes of living, patterns of exposure and vulnerability of specific social groups. They not only operate in the individual domain but also in the general domain of society and in the particular domain of clearly differentiated social groups.

This transformative dynamic interpretation of socially determined critical processes that produce health estates implies a deep interpretative innovation which departs from the linear logic of risk factor causality. It also expands the scale of prevention and promotion to social processes that operate beyond the individual domain, transcending the conventional risk factor model of conventional Cartesian epidemiology.

Methodological Ruptures

The "knowledge illusion" of linear, reductionist thinking lies in its capacity to build consistent formal models that describe variable associations, correlate them and predict their formal links and behavior under defined probabilistic conditions.

This Cartesian deterministic logic assumes the World as a set of parts that are linearly connected. In epidemiology, those parts are the "risk factors". In the more elaborate formal models, those risk factors are arranged according to a specific conceptual model. For example: physical environment factors; biological environment factors; social environment factors. This kind of formal arrangement was applied in the natural history of disease theory of the Leavell-Clark's ecological triad model; an approach that came to be well known for introducing the concepts of primary, secondary and tertiary prevention (Cockburn 1963; Leavell and Clark 1965).

Under the empirical linear framework, a typical affirmation would establish, for example:

> The risk of type 2 diabetes is determined by an interplay of genetic and metabolic factors. Ethnicity, family history of diabetes, and previous gestational diabetes combine with older age, overweight and obesity, unhealthy diet, physical inactivity and smoking to increase risk. (WHO, p. 12)

These are unquestionably real empirical components of the diabetogenic process, but they merely configure a set of external connections of parts. I will discuss further about the interpretative flaw that arises when we substitute a real explanatory model of a certain movement in the epidemiological process of determination with the formal linear representation of the conjunctions between its parts (variables), albeit without an explanation of their contextual relations and concrete historical interplay.

In these lines, the main argument is that movement, as conceived within social determination, is not a Cartesian part-to-whole determinism, but rather a dialectic process where the totality-to-part movement constitutes a two-way dynamic articulation.

Going back to our previous example on diabetes, the empirical linear description makes invisible the historical transformation of modes of living and the global dissemination of what has been aptly called the "neoliberal diet". This explains, among other things, the social class differentials in obesity between high-quality food consumption in the upper classes and energy-dense fares for the poor in USA (Otero et al. 2015).

In this explanatory approach, the argument on diabetes could be restated:

> Diabetes, a major mode of living disorder, has become a global burden, and the prevalence rates are rising steeply in developing economies. Rapid socioeconomic transition with urbanization and industrialization are the main causes for the global diabetes epidemic. Among developing economies, the highest increase in number of people with diabetes is in China followed by India. (Pradeepa and Mohan 2017)

As stated before, conventional empirical linear epidemiology harbors a number of methodological and epistemological assumptions about the society it seeks to analyze. These assumptions are often implicit within the discipline and are rarely presented as one methodological or epistemological orientation among many. In

Gramscian terms, they have attained a hegemonic or "common sense" status within epidemiology specifically and the health sciences more generally. Some of these basic assumptions will be dealt with in this section.

Latin American critical epidemiology and its social determination paradigm are a historical result. Its new ideas, methodological and practical contributions are preceded by a series of conceptual ruptures or breaks with the conservative functionalist paradigms of each historical period. In the Colonial period, progressive miasma theory supporters confronted the contagionist logic. Then, in the first half of the twentieth century, the progressive contributions of the forefathers of social medicine opposed the reductionist explanations of positivists. In the second half of the twentieth century, the main interpretative dissent occurred between the empirical multicausal ecological model on one side, and the progressive views of authors of new social medicine and critical epidemiologists on the other. Currently, in the twenty-first century, the main scientific dissent has occurred between the social determinants perspective—its mono-cultural and institutionally based perspective—on the conservative side, and critical epidemiology on the progressive side.

Throughout these periods, the principal object of dissent has been the claim by progressives that conservative approaches to epidemiology fragment reality and divorce the analysis of specific health problems from their historical context. These ruptures also reflect the concern that, if epidemiology is to become emancipatory, it requires not only an ethical revolution but theoretical, epistemological, and methodological transformations as well. Below I distill this process into five distinct but interrelated ruptures.

1. Causal Linear Thinking Versus Complex Social Determination

The explanatory power of science is a potent tool for social control. Be it for practical productive purposes or for ideological reasons, knowledge is key to the construction of hegemony. This characteristic has placed scientific work, as I argued before, under the permanent pressure of economic and political power. Under those circumstances, dominant nation states have institutionalized a form of official science convenient to the imposition of functional research and a methodology that—sometimes willingly, sometimes not—ends up supporting an ordering of the world according to prevailing conditions of acceptability (Foucault 1982, pp. 208–226). Contrarily, the strategic needs of the people demand independent critical thinking and transformative science.

Scientific empirical reductionism separates real events into things, on the one hand, and their "contexts and evaluative relations", on the other (Irvine et al. 1979). This is a philosophical distinction that allows for a subjective manipulation of facts. Reductionist thinking, as explained further below, is the interpretative model that fragments the world in parts and concedes preeminence of the parts over the whole. This philosophy first appeared in astronomy and physics and later in physiology and biology, where it established a set of false separations (Irvine et al. 1979, p. 66), such as ones between:

subject–object
purpose–mechanism
value–fact

internal–external
secondary–primary (properties)
thought–extension
mind–body
culture–nature
society–science

In order to subordinate scientific interpretations to empiricist rules of objectivity (i.e. the notions of matter, motion, and number), positivist science subsumed the concepts of the first column in their corresponding ones from the second group. In other words, positivists explained "the concepts and phenomena of the first column by those in the second" (Irvine et al. 1979, p. 66). From an epidemiological perspective, we can add three more conceptual substitutions that empirical epidemiology has implemented to substitute complex interpretation of health reality with its linear description:

explanation–prediction
processes–factors
determination–causes

This first rupture entails a break with the interrelated notions of causality and linearity and an embrace of determination and complexity. I will not delve here in the entire history of how linear epidemiology was challenged by various advocates of the new Latin American epidemiological thought from the 1970s through the 1990s—authors like Laurell (1978, 1982), Samaja (1993), Donnangelo (1979, 2014), Almeida (1989, 2000), Tambelini (1978), Menéndez (1981, 1998, 2008), Ayres (1997), Victora (1992) and myself—all of whom provided important contributions toward a refounding of modern critical epidemiology. These works have been widely commented on and documented (Duarte 1986; Franco et al. 1991; Waitzkin et al. 2001), and I have summarized them in the chapter "Latin American Critical Epidemiology", which is part of the latest edition of *Epidemiology: Political Economy and Health* (Breilh 2010). Suffice to say that linearity is the organizing temporal-spatial metaphor of empirical epidemiological inquiry. Its use implies accepting both a linear order regarding disease generation in conjuction[1] with various "risk factors" (individual, behavioral, cultural, social, and even structural) and their biological effect in individuals. This linearity presents a fragmented understanding of social reality (linearly related and ultimately decontextualized, where risk factors ultimately have biological impacts) and a notion of the individual who is susceptible to these risk factors as standing apart from that social reality and nature. The social determination of health paradigm, on the other hand, substitutes causal linearity that predicts the individual health effects of so-called "risk factors" with complex social processes that explain the production of collective health conditions, which in turn explain familiar and personal conditions. This paradigm

[1]Conjunction refers to external causal links; it is fully discussed further on.

treats individuals not as separate from the social reality these processes give rise to, but rather as constitutively produced by these social processes and, at the same time, as agents in shaping them.

Those reified "risk factors," which are actually subtracted from the complex social processes they make part of, are separated from their contextual socio-economic conditions and treated as causal determinants. Rather than this risk factor-based notion of causality, we propose the notion of determination, which entails analysis of the relation between pathology and the social reproduction movement of society as a whole. The social reproduction of market society is dependent on the processes of capital accumulation (structured wealth expansion and concentration). This whole social movement generates the subsumption of the natural ecosystems and of the human biopsychological transformations in the movement of economy, politics and cultural hegemony. Private capital accumulation becomes the rule and logic of the whole society and its regions; it becomes the norm of distribution of traits and benefits and ends up submitting all natural and social processes to its logic. This form of social reproduction generates and imposes modes of living of different social groups (i.e. under social class, gender and ethnocultural relations); living modes which imply work and consumption patterns; organizational capacities and collective forms of support; cultural characteristics and forms of identity; and specific relations with the ecosystem. These structured modes of living subject those social groups to healthy, protective conditions and at the same time to unhealthy, destructive ones. Here, a dialectical dynamic presupposes typical patterns of exposure to those favorable and unfavorable processes and also generates typical forms/patterns of vulnerability to health hazards. The terminal expression of this contradictory movement is the embodiment in the concrete phenotypes and genotypes of biopsychological traits. The explanation of these processes surmounts and transcends the limited, fragmented, and reified linear-causal logic that fragments this complex movement by separating the so-called "risk factors" from underlying processes.

Also from this perspective, the limitations of the seemingly progressive social determinants of health theory become evident. It is very important to distinguish the substantial ontological, epistemological and practical differences between *social determination* and *social determinants* of health. The determinants of health applied to particular populations remain formally sound abstractions that are fundamentally emptied of their critical content through their situatedness within the empirical linear-causal paradigm and their systematic disseverance of these deeper processes of determination which ultimately shape epidemiological patterning. I should point out, however, that *determination* is not synonymous with *determinism*, as I have explained previously (Breilh 2003). *Determinism* refers to the philosophical doctrine arguing that any type of event is connected to a cause and that, given an effective cause, the effect invariably occurs. On the other hand, *determination* argues that all processes gain their essential characteristics within a concrete historical form of movement. This mode of movement is the mode of becoming, of gaining its properties. The difference just referred implies another important contrast: while causal relations are merely predictive, social determination processes are generative or productive (i.e. explanatory). The contrast between prediction and explanation lies, as mentioned, at the basis of the difference between functional and critical epidemiologies. Finally, while social determination calls for an integral, transformative action that targets the whole multidimensional reality of health, the causal deterministic fragmenting perspective is confined to focalized actions on causal risk factors.

2. Conjunction Versus Subsumption

In classical linear-causal epidemiology, causation is ~~expressed~~ based on the principle of conjunction, which was first elaborated by Hume and which was later systematized by Bradford Hill (1965) in his memorable presentation on association and causation. As I pointed out in my first work on the social determination of health (1977), Hume's notion of causality prevailed as the guiding logic of epidemiology, a fact that was explicitly recognized in the main hegemonic epidemiology texts of the 1970s (MacMahon 1975) and is still recognized in more recent texts (Rothman 2008). Hume's notion of conjunction stipulated that the efficient operation of a cause results from its external connections with an effect, which is what affects individual phenomenal effects. In this case, the ruling principle is causation by conjunction or external connection with factors of a reified and fragmented environment (MacMahon 1975; Rothman 2008). This fragmented conception of the social order leads to the conclusion that the action of public health is equally causal as regards risk factors. In the place of Humean conjunction as the cause-effect logic of epidemiology, I proposed the concept of subsumption, which was first presented in a series by PAHO/WHO (Breilh 1994).

Subsumption refers to the ways in which simpler processes taking place in the individual, psychological, and biological (phenotypic and genotypic) realm are shaped (or subsumed) by more complex social processes. This implies a continuous, structured conditioning of the simpler processes by more complex ones and opens the door to analyses of, for example, the determination of epigenetics and genetic expression under advanced capital accumulation. These are not simply external relationships between distinct systems, but ones that together form a concatenated social whole that constitutes the social system.

Material organization of life is transformed along with the ways of thinking and political organization. In that movement, there is a subsumption of the less complex processes in the more complex ones.

In our interpretative model of critical epidemiology, we underline the fact that concrete material life and eventually health are determined through forms of subsumption (Breilh 1977, 2003, 2010). That means that individual phenotype and genotype phenomena are conditioned by social relations (Levins and Lewontin 1985), which are determined in the general societal domain and in the particular domain of groups with their social power relations and collective modes of living (i.e. class, gender, ethnic). This movement finally is incarnated or expressed in forms of embodiment in individual bodies (Krieger 2011).

The difference between *conjunction* (i.e. external causal links) and *subsumption* (i.e. inherent connection) is of paramount importance to the transformation of epidemiological reasoning; it also has important contrasting implications for action. Substituting causality for subsumption commits epidemiologists to embrace and achieve transformative health science; it is the only legitimate pathway to radical measures expected instead of mere functionalist factor modifications.

So, why are we arguing in favor of this conceptual-practical shift? It is precisely because of the transformative logic that subsumption makes possible as it:

- Links inherently specific processes and variables to the collective movement of society and its particular settings.
- Connects the material or structural base and organization of society with its thinking, political and cultural forms.
- Understands simpler processes in the framework of the more complex ones. That is the case of biological natural processes of humans and ecosystems being dialectically subsumed in the historical social processes.
- Explains that individual, biological human and ecosystem forms are subsumed under social processes, withholding spaces of relative autonomy.
- Integrates the notion of social space as a socially determined but at the same time socially productive element with the dialectical conception of social determination of health; superseding the empirical health geography model.

Subsumption is a category that has evolved to explain different dimensions of social movement, which we have systematized (Breilh 2015):

- Formal subsumption in manufacture and real subsumption in industrialized work (Marx 1971; 1972: nineteenth century)
- Biological subsumption in the social (Levins and Lewontin 1985; Laurell and Noriega 1989; Samaja 1993; Samaja 1996; Breilh 1977, 2003, 2010; Krieger 2011: 2nd half of the twentieth century)
- Subsumption in consumption (Veraza 2008; Barreda 2010: twenty-first century)
- Cybernetic subsumption (Breilh 2015b: twenty-first century)

3. Empirical Statistics and Qualitative Observation Versus Critical Statistics and Qualitative Observation

Modern society placed quantification at the center of positivist science. By the late fifteenth century, positivists defined quantitative measurements and relations as: *what counts in knowledge*. But to free those measurable elements from their social connections, they had to be separated from their qualitative historical evaluative relations. This logical separation was crucial for the sake of manipulating, dominating, and exploiting both natural processes and social relationships; it was of greatest importance to transforming socially generated data into useful practical decontextualized abstractions. In logical terms, Cartesian statistics also supported a type of social science concerned with prediction instead of explanation, convenient to the reproduction of the blossoming capitalist society (Leiss 1972). Thus, through reifying a logic of quantification, decontextualization and prediction, dominant Cartesian research and statistics were positioned as a means of pragmatic manipulation and commodification of natural and social objects (i.e. epidemiological objects), instead of a critical transformative explanation of reality.

In order to consolidate this model, Cartesian science applied reduction as the method and reductionism as its theoretical perspective. Reductionist analyses explain the properties of complex wholes in terms of the units of which those wholes are comprised. In other words,

> reductionism is the claim that the compositional units of a whole are ontologically prior to the whole that the units comprise. That is to say, the units and their properties exist before the whole, and there is a chain of causation that runs from the units to the whole…social phenomena are the sums of the behaviors of individuals. (Lewontin et al. 1984, pp. 5–7)

From this perspective, the world comes to be perceived as a convergence of empirical particles, with the underlying principle being that it is the properties of those particles that explains and determines the whole. From an epidemiological perspective, it is also important to emphasize that reductionism recognizes "true causes" as arising only in the empirical plane, therefore neglecting the ontological importance of the social context (Levins and Lewontin 1985).

Adhering to this interpretative paradigm, as I have argued before, conventional empiricist epidemiological statistics build explanations and data patterns about a population's health under the assumption that it is the variations and relations in the "peak of the iceberg" of empirical reality, of those particles and their probable conjunction, that provide statistical evidence of the whole (collective health). They explore statistical regularities under the ontological premise that the whole is a result of the parts and that collective statistics are the summary of individual observations and their system of relations. Contrarily, critical science has exposed the reductionist logic of empirical Cartesian statistics that considers all phenomena "as explicable by deterministic cause and effect at an arbitrary level of explanation" (Levins and Lewontin 1985, p. 153).

Bhaskar (1986) has clearly explained the sequence of reductions that end up separating empirical evidence from reality and the conceptual reductions that convert the experimental logic as the core framework of a supposedly objective but in fact truncated science. He explains the sequence of methodological reduction. The first step is to decontextualize empirical data patterns from their social context, working only on empirically observable conditions (patterns) and leaving out non-empirical processes and contextual determination. Through this operation, the generative mechanisms of movement are left aside and only event patterns are instated. Then, from those event patterns, non-associative movement patterns are left out, keeping in the analysis only present constant conjunction variables. And finally, in the third stage of reduction, only conjunctions that comply with experimental closed system rules are kept and become causal prescriptions.

Therefore, as I have stated, reductionist logic presupposes causality as the sole condition of epidemiological explanation and as the rule for the generation of health conditions. Contrarily, critical epidemiology questions this linear, reductionist and static factorial conception of causal thinking, substitutes it with the complex dialectical idea of process determination, and thus redefines the whole-to-part relation. But in so doing, it needs to innovate the theoretical basis of Cartesian statistics. This innovation demands a new statistical model which implies the need to reconsider conventional Cartesian resources in order to recuperate the whole-to-part relation in the generation of health conditions. This is important to cover the different forms of movement—not necessarily Cartesian—that this integral process implies. In other words, we need a statistical theoretical and operational framework that allows us to: (a) approach different forms of movement that characterize epidemiological processes in the general (society), particular (groups) and singular (individuals) domains of society; (b) redefine the role of probability and the relation between stochasticity and determination; and (c) elaborate new interpretations of measurement, classification, and comparison in order to understand movement or variability in all its forms.

In order to advance in that direction, we must take into account the fact that whatever theoretical perspective it supports, epidemiology constantly needs to incorporate variables in its analysis of empirical phenomena. When doing so, it needs to assemble a variables system, defining the supposed space, trends and patterns of variation. By adopting the Cartesian principles, conventional epidemiology statistics logically assume variables as expressions of the variations that occur in individuals and individual risks that move in a regular dynamic system, be it bivariate or multivariate. Under this theoretical and methodological perspective, differential equations and data mappings are constructed of individual variations, assuming them as the rules of movement and relations. It makes decontextualized arrangements of the so-called independent and dependent variables that interplay in the system. From this perspective, statistical analysis mainly takes form through contingencies, correlations, variance analyses and factorial groupings of reduced empirical 'peak of the iceberg' expressions (Breilh 1997).

Research done through this theoretical approach has made important contributions to the understanding of empirical epidemiological variable analysis, explaining the isolated variability or *movement of cause-effect conjunction* in effective causal relations. In fact, the statistical demonstration of quasi-experimental causal relations has accumulated an abundant valuable bibliography. However, critical epidemiology needs to move beyond this reductive understanding of variability or dynamicity by expanding statistical management of quantitative data, not only to recontextualize causal analysis as part of the social determination process but to focus and integrate new statistical constructions. From this perspective, it is indispensable to apply the conventional tools of statistics in connection with the social determination processes, and also to develop other tools for studying quantitative variability. This in order to uncover other forms of movement in the explanatory model: the *movement of contradiction* between healthy (i.e. life supportive and protective) and unhealthy (i.e. life vulnerability and destructive) processes that operate in the social and individual level; the *feedback movement* of embedded systems; the *stochastic movement* that operates in dynamic regular systems under relative limits or degrees of freedom, not only determined internally at the individual level—whether particles or individuals—but by subsumption under the conditions of a higher level, as will be explained (Breilh 1997, 2003). And, finally, in some moments and spaces, epidemiologic phenomena move under irregular dynamic non-linear systems that result under certain initial conditions which trigger exponential uncertainty. This corresponds to *chaotic movement* (fractal behavior), which tends to evolve toward specific numerical values referred to as attractors (Osorio 1988; De Castro Ildeu 1992).

The importance of considering this multidimensional conception of movement is double. First, it expands the notion of epidemiological movement or variability from variance—restricted to the stochastic movement through causal conjunction of independent and dependent variables—to the other forms of movement just mentioned. And second, it makes possible the conversion of quantitative analysis, from a rigid tool restricted to the quasi experimental observation of linear relations (conjunctions) of empirical indicators (risk factors and outcomes), to a dynamic resource that is capable of linking empirical evidence to its dynamic determinant context. Whatever the complexity of any linear multidimensional and multi-criteria analysis model applied, empiricist designs restrict their epidemiologic hypothesis and conjectures to the logic of linear causality and reduce their scope to the associations and correlations between the so-called “risks factors” and their “outcomes”, while restricting both to the probabilities of empirical phenomena that occur in the peak of the iceberg of social determination. It is necessary to inscribe these analytic tools in the dialectic movement of epidemiological determination and subsumption occurring in the general, particular, and singular dimensions of social reproduction.

As discussed, critical epidemiology overcomes the logic of causality with the logic of complex determination of health. For this, it must redefine the relation between stochastic and deterministic processes and also the role of probability in analysis:

> ...there is no reason to make uncertainty an ontological property of all phenomena... Not only can the apparently random be explained as a result of deterministic forces in higher dimensionality with more specification, but a reduction in dimensionality by averaging also converts stochasticity into determination. (Levins and Lewontin 1985, p. 153)

In the face of all arguments put forward here, conventional statistics, in the first place, do not take real variation as their object of study but, instead, use a constructed variation, largely by reducing, discounting and separating processes of other domains of social life, considering these as “noise”. This maneuver divorces the constructed evidence from such reality. In this way, the conventional understanding of variances through the analysis of variance, multivariate correlation and regression becomes a reifying procedure in consonance with the Cartesian logic that assumes such correlations as equivalent to the real world. As Levins and Lewontin further state,

> Correlation can create relationships that do not exist...the correlation between a pair of variables, even in the simplest ecosystems, depends, first of all, on the rest of the structure of the system; second, on the variable at which the external source of variation enters the system; third, on the history of the system; and, finally, on the duration of the observation. (1985, p. 158)

Thus, the notion that a realist analysis comes from the strict application of statistics is inconsistent both in biological and social sciences. “Some of the great problems of scientific explanation come from concepts and practices that lie in the heart of modern statistics, which are in many ways the embodiment of idealism” (1985, p. 154). That is of course the logic of Cartesian statistics.

A last important argument in this regard is that not only do we need to surpass linear reductionist understanding of variability, but also to overcome the social classification systems that Cartesian logic applies. Classifications, being a "spatial or spatio-temporal segmentation of reality. . .when they become visible, [they] become objects of contention" (Bowker and Star 1999). In epidemiology, when we classify the population by placing them in "boxes", we are applying—knowingly or not—a social classification system that implies certain meanings, identities, and—of course—a Cartesian segregation of our observed cases. This conduces only to individualized health actions. Reductionist classifications are based—as explained—on decontextualized individual traits.

Qualitative observation: also in need of critical valuation

Methodology depends on the object of study. It is structured and develops around the object's characteristics and movement. Real phenomena are not only characterized by their quantitative aspects but also by their attributes. So, real processes possess quantities in relation to given qualitative attributes. Qualitative and quantitative development is interdependent and, to understand the essential traits of measurable evidences, we cannot disconnect them from their qualitative historical evaluative relations. Quantitative and qualitative attributes of our population and its health are not essentially individual creations but are formed and transformed under complex social determination (Breilh 1997).

For the same reasons that we have contested the preeminence of the quantitatively measurable as the only source of what counts in science, we must be clear that the solution to mono-methodic positivist quantitativism is not replacing it with rationalist mono-methodic approaches that operate within a form of qualitativism. The spread of qualitative approaches built from phenomenological and constructivist frameworks, the appearance of computer-based qualitative analysis and its powerful software introduced positive pressures on the development of mixed quali-quanti analysis (Tashakkori 1998; Punch 2014).

Epidemiology as a socio-biological science requires understanding complex systems. It therefore needs to incorporate complex thinking to explain the actual material relations and contradictions between healthy, supportive and protective processes, which are affected or contradicted by unhealthy, hazardous processes.

As I argued previously, three dimensions make part of this intricate multidimensional process of social determination of health. First, the broader dimension of social reproduction in a society as a whole, with its economic movement, its corresponding cultural and political power relations. Then, the particular dimension of social classes, crossed by gender and ethnic power relations, that govern the appearance of specific modes of living which account for their specific modes of social reproduction. Thirdly, the singular dimension of individual and family styles of living that are made possible by the other two dimensions. Dialectically speaking, even if the more complex and comprehensive processes subsume the movement of the lesser ones—i.e. the general over the particular and this last one over the singular—there is a relative autonomy in the variation of the less complex which allows for a relative autonomy of the singular with respect to the particular and the particular with respect to the general processes (Breilh 1977, 2003).

In such multifaceted complex whole, a contradictory interplay develops between the tendency of the whole system to reproduce itself, conserving its defining characteristics, on one side, and, on the other side, the tendency of its parts to generate changes (Samaja 1996).

The dialectic thrust of this movement implies the opposition of subsumption tendencies that submit the particular groups to the broader logic of general social reproduction conditions, and the singular individual living styles to the broader logic of their classes' mode of living. But, at the same time, the contrary relative autonomic movement of individuals in relation to their groups and of groups in relation to their society as a whole is the essential trait of the permanent transformation of epidemiological conditions.

In our Research Center of Social Impacts on Collective Health (CILAB*Salud*) and in our doctoral and postdoctoral Programs on Collective Health (Universidad Andina Simón Bolívar -Quito, Ecuador-) we are working on the innovation of statistic models and resources for epidemiological research.

4. Cartesian Space Versus Socio-historical Space: Reframing Health Geography

The epistemological debate proponed along these pages becomes especially evident and necessary when we analyze the role of one of epidemiology's basic tools: health geography. Here, as well, there is a conceptual clash between Cartesian empiricist and critical science.

From the Cartesian perspective, the role reserved for geography in epidemiological studies is that of an external "place" (i.e. map) where physical, social and environmental health "factors" and population can be located. Either in conventional empiricist cartography or in modern digital geographic information systems, the empiricist perspective reduces geographical space to a passive container where natural, economic, cultural or social phenomena are displayed. The landscape is taken for a collection of natural accidents, animals and forms of vegetation.

Broadly speaking, Cartesian geography constructs in various layers: places (locations), distances, pathways of mobility and spatial organization. Geographical constrictions are no more than spatial relations of empirical objects. Geography is merely a spatial classification tool. Space is solely a mental metaphor for empirical situations.

The dualism of Cartesian logic is characterized by systematic and distorting separations or dichotomies. As we mentioned before, it creates an artificial epistemological divide between subject and object, matter and spirit, the internal and external, the material world and the mind, culture and nature, among others. In geography, it separates geography and history, it creates a false divide between geographic space and its historical context.

In the background of this epistemological debate resides the conceptual and ideological dispute around the notion of space. In order to recuperate the objective dynamism of the social space, not as sole receptacle in which social life and epidemiological phenomena develop, we must free this notion from the masquerading reductionist reasoning of linear empiricist thinking.

The fundaments for a renewed critical geography were established by Lefebvre (Lefebvre 1991; Harvey 2007; Santos 1996).

Geographical space, as one of the forms of social space, is not a passive receptacle. The social space produces society and at the same time is produced by social relations. Therefore, to understand the geography of social phenomena (i.e. epidemiological processes), we must recognize that it definitely participates in their social determination, that there is a two-way determination between society and its social space.

> Can space then be nothing more than the passive locus of social relations, the medium in which their combination is transformed, or the aggregate of the procedures used in their dismantling? The answer is no, I will show later that active role, operative, instrumental space, as knowledge and action, in the existing mode of production. (Lefebvre 1991, p. 11)

There is a historical evolution of the social space that generates all forms of space (including the geographical space) with their connections, networks, relationships and dynamics. Social spaces interpenetrate and superimpose each other—a notion that supersedes the abstract view of classical mathematics (Euclidean/Cartesian).

Thus, social space participates in the material condition of human and natural forms of life. When one analyzes, for example, social inequity, we immediately think of social power relations. It is nevertheless important to realize that "inequality is not only constructed and reproduced through the inequalities of work and market, but through the overwhelming penetration of a material organization of life, of a model to organize the practice of living" (Breilh 2011). In that case, the intermediating role of spatial determination is fundamental.

To finalize these epistemological and methodological reflections on space and health and on the relation between space and power, I must bring into consideration here the paramount importance of cyberspace in the social determination of health.

Three years ago, I proposed the epidemiological use of two categories: cyberspace and cyberdetermination. It was my keynote speech for a conference on the critical geography of health, later published in the *Brazilian Journal of Epidemiology* (Breilh 2015b).

I was motivated at the time by the Snowden affair. Our research group was especially concerned about the dominating penetration of economic and political power in our inner intimate life. And in doing so, I predicted the overwhelming and fast growing social and health impacts and proposed the notion of cybernetic subsumption.

No matter from what theoretical-epistemological or ideological-political horizon we approach the present crisis of natural and human life, we have to recognize the advent of new forms of virtual or cybernetic unhealthy subordination of human life. These not only include economical exploitation and domination by force, but new subtle forms of hegemony and social subsumption—most effective and less obvious—that are reproduced in the cybernetic domain.

My main argument was that, in the last decade, the dominance of big corporations in cyberspace has become visible as a new historical process that conditions the social order and extends the technological subordination of daily life, therefore expanding community massive submission to standard conducts and powerful interests. The new digital technological revolution, about which some frightening prognoses have been made, could easily imply the advent of an era of radical subsumption of life processes. This will negatively affect not only our general way of living, thinking and planning, but also our deepest daily intimacy. This movement implies radical effects on health which we call cybernetic determination and subsumption. This novel process raises new questions on public health and prevention; but it also requires a new reading of reality, a rethinking of human life and health, of its social determination. And this entails the need for new categories and analyses and renewed challenges for critical epidemiology.

Critical Epidemiology and the Critique of Functional Public Health

Prevailing functional public health is ~~entramped~~ hobbled by formal planning schemes, usually restricted to compliance with quantitative institutional performance and the reduction of prioritized epidemiologic rates. Although it is perfectly understandable that the health system needs to put forward concrete assessable action goals, critical epidemiology nonetheless goes beyond periodic focalized factor-correcting objectives, and embraces integral, intersectoral, transdisciplinary and intercultural programs. In order to make real contributions to communities' wellness, it builds action around the participative construction of optimal collective and personal life standards.

When it comes to defining what fundamental principles should inspire and guide a renewed public health (i.e. collective health in Latin American terms) that have to be taken into consideration when evaluating the health estate of a society, of a region or of a specific territory, our approach adopts an evaluation system that I have called: the four "Ss" of life.

- A sustainable economy for the reproduction of living beings;
- A sovereign libertarian identity and culture;
- Solidary politics and agency for equitable relations;
- Integral human and natural bio-Security, only feasible in communities that have built non-toxic, clean, society-nature metabolism relations and healthy, safe environments in the workplace, in the neighborhood; safe forms of transportation and non-restricted access to health and wellbeing public services.

In these lines, epidemiology must pay special attention to the degree to which those requisites or wellbeing goals are attained. Unfortunately, those four basic conditions for wellness and healthy living are systematically destroyed and structurally denied. Short sighted big business severely jeopardizes sustainability. We work, eat, rest and live impregnated and surrounded by "poisons", toxic and carcinogenic, and at the same time wasting, draining and sequestering vital resources. Sovereignty—either personal, regional or national—is systematically destroyed by the forceful or elegant imposition of foreign profit interest. Every day the capacity for personal or collective self-determination suffers serious drawbacks through economic, legal, cultural violent mechanisms. People lose control of their vital resources, and local or national decisions increasingly depend on unfair economic or political negotiations. Solidarity is heavily compromised by overwhelming individualism and radical consumerism. We experience the extensive and continuous denial of property and rights of life in equity. Bio-Security is threatened severely and rampantly by the growth and multiplication of unhealthy processes and spaces; by the exponential growth of dangerous garbage.

In other words, responsible conscious epidemiological research and teaching cannot ignore the fact that big business has placed humanity at the edge of a social an environmental abyss. This is by no means a magnification of factual truth.

The productive force and leitmotiv of the global entrepreneurial system is aimed exclusively at the generation of profit from commodities (objects that contain accumulated labor, "dead work"), instead of being focused on the production of conditions for a plentiful subsistence of living beings. It relies on an unsustainable, inefficient, degrading and wasteful form of productivity. It submits communities to dependent, alienating and subordinate modes of living. It destroys solidary collective-based relations by undermining material and cultural life supports, by replacing ancestral solidary cultures and habits with self-centered behavior patterns, as part of a culture of individual survival in the midst of a civilization of loneliness. And finally, it converts Nature in a source of profitable commodities: water, seeds, plants, animals, the microbiota, genomes, geophysical forces, etc. From this destructive and unethical perspective, human health, security and rights of workers and their families are only a strategic burden to be tackled.

There is an urgent need to redirect research and teaching toward the understanding of this global health crisis; work directed not only to its unveiling but to build a new project of civilization. We urgently need to expand critical knowledge, technical instruments and tools for accountability, and social control over the economic, political and cultural policies that affect the four "Ss" of life.

We must strive to dismantle in our universities and health entities the hegemony of the pharma-bio-medical model. Our struggle in this direction will not only benefit social medicine/collective health but also the development of independent basic health science and clinical knowledge. It is not only social and academic vindications but human survival that are at stake.

Final Conclusion

If we assume critical, innovative thinking as the reference for integral scientific objectiveness, and if we are conscious of the complexity of social life and the ethical implications of health actions, we can decipher the real requisites of rigorous epidemiology, which do not only depend on the accuracy, reliability and validity of its empirical methods, calculations and observations but on a rigorous and multidimensional/complex/transdisciplinary/intercultural perspective. This involves the understanding of how a redefined ethical-political role for epidemiological research demands new criteria of objectiveness and of the research subject in defining and explaining the health situation integrally, without the ambiguity and empirical reductions that underlie quantitativist empirical reasoning.

References

Allende, S. (1939). *La realidad medico social chilena*. Santiago: Ministerio de Salud.

Almeida Filho, N. (2000). *La ciencia tímida: ensayos de de deconstrucción de la epidemiología*. Buenos Aires: Lugar Editorial.

Ayres, J. R. d. C. M. (1997). *Sobre o risco: para compreender a epidemiologia*. São Paulo: Hucitec.

Barreda, A. (2010). *Economía ecológica y ecología crítica*. Quito: Seminario del doctorado en Salud Colectiva Ambiente y Sociedad, Universidad Andina Simón Bolívar, Sede Ecuador.

Bartra, A. (2009). *El capital en su laberinto*. México, DF: Universidad Autónoma de la Ciudad de México: Editorial Itaca: CEDRSSA (Mécico Bartra, A. (2006). *El capital en su laberinto: de la renta de la tierra a la renta de la vida*).

Bhaskar, R. (1986). *Scientific realism and human emancipation*. London: Verso.

Bourdieu, P. (1998). *O poder simbólico* (p. 15). Rio de Janeiro: Bertrand Brasil.

Bowker, G. C., & Star, S. L. (1999). *Sorting things out: Classification and its consequences*. Cambridge, MA: MIT Press.

Bradford Hill, A. (1965). The environment and disease. Association or causation. *Proceedings of the Royal Society of Medicine, 58*, 295–300.

Breilh, J. (1977). *Crítica a la interpretación ecológico funcionalista de la epidemiología: un ensayo de desmitificación del proceso salud enfermedad*. México: Universidad Autónoma Metropolitana de Xochimilco, Tesis de Maestría.

Breilh, J. (1994). Las ciencias de la salud pública en la construcción de una prevención profunda. In M. I. Rodríguez (Ed.), *Lo biológico y lo social: Su articulación en la formación del personal de salud* (Serie Desarrollo de Recursos Humanos) (Vol. 101, pp. 63–100). Washington: OMS.

Breilh, J. (1997). *Nuevos conceptos y técnicas de investigación. Guía pedagógica para un taller de metodología* (3ra. Edición, Vols. 1–1). Quito - Ecuador: CEAS.

Breilh, J. (2003). *Epidemiología crítica ciencia emancipadora e interculturalidad* (2da ed.). Buenos Aires: Lugar Editorial.

Breilh, J. (2010). *Epidemiología: economía política y salud* (7a ed.). Quito [Ecuador]: Universidad Andina Simón Bolívar/Corporación Editora Nacional.

Breilh, J. (2011). The subversion of the good life (enlightened rebelliousness for the 21st century: A critical perspective on the work of Bolívar Echeverría). *Salud Colectiva English Edition, 7*(3), 389–397.

Breilh, J. (2015). Epidemiología crítica latinoamericana: raíces, desarrollos recientes y ruptura metodológica. (La determinación social de la salud como herramienta de ruptura hacia la nueva salud pública – Salud Colectiva). In *Tras las huellas de la determinación (Memorias de Seminario Inter-universitario de determinación social de la salud)* (pp. 19–75). Bogotá: Universidad Nacional de Colombia.

Breilh, J. (2015b). Epidemiologia del siglo XXI y ciberespacio: repensar la teoría del poder y la determinación social de la salud. *Revista Brasileira de Epidemiologia, 18*(4), 965–975.

Breilh, J., & Tillería, Y. (2008). *Global acceleration and dispossession: Regression in the right to life and public health during Ecuador's neoliberal decades* (1st ed.). Ottawa: University of Ottawa.

Carrillo, R. (1951). *Plan sintético de Salud Pública 1952 1958*. Buenos Aires: Ministerio de Salud Pública de la Nación.

Cockburn, A. (1963). *The evolution and eradication of diseases*. Baltimore: The Johns Hopkins Press.

De Castro Ildeu, G. (1992). Caos. Rio de Janeiro: Ciência Hoje. *Sociedad Brasileira para o Progresso da Ciência (UFRJ), 14*(80), 62–63.

Duarte, E. (1986). *Ciencias sociales y salud en América Latina: tendencias y perspectivas*. Washington: Organización Panamericana de la Salud.

Espejo, E. (1994). *Reflexiones sobre la utilidad, importancia y conveniencias que propone Don Francisco Gil, Cirujano del Real Monasterio de San Lorenzo, y su sitio, e individuo de la Real Academia Médica de Madrid, en su disertación físico-médica, acerca de un método seguro para preservar a los pueblos de viruelas*. Quito: edición facsimilar de la Nueva Editorial de la Casa de la Cultura. (Facsimilar). Quito: Nueva Editorial de la Casa de la Cultura Ecuatoriana.

Foucault, M. (1982). The subject and power. In H. L. Dreyfus & P. Rabinow (Eds.), *Michel Foucault. Beyond structuralism and hermeneutics*. New York: Harvester Wheatsheaf.

Foucault, M., Lotringer, S., & Hochroth, L. (2007). *The politics of truth*. Los Angeles, CA: Semiotext(e).

Franco, S., Duarte, E., Breilh, J., & Laurell, A. C. (1991). *Debates en medicina social* (Desarrollo de Recursos Humanos) (Vol. 92). Quito: OPS.

Harvey, D. (2003). *The new imperialism*. Oxford: The Oxford University Press.

Harvey, D. (2007). *Espacios del capital: hacia una geografía crítica*. Madrid: Akal Ediciones, S.A.

Irvine, J., Miles, I., & Evans, J. (Eds.). (1979). *Demystifying social statistics*. London: Pluto Press.

Jones, R., & Wilsdon, J. (2018). *The Biomedical Bubble. Why UK research and innovation needs a greater diversity of priorities, politics, places and people*. London: Nesta.

Klein, N. (2008). *La doctrina del shock. El auge del capitalismo del desastre*. Buenos Aires: PAIDOS.

Krieger, N. (2011). *Epidemiology and the people's health: Theory and context*. New York: Oxford University Press.

Laurell, C., & Noriega, M. (1989). México: ERA.

Leavell, H., & Clark, G. (1965). *Preventive medicine for the doctor in his community*. New York: McGraw-Hill.

Lefebvre, H. (1991). *The production of space*. Malden, MA: Blackwell.

Leiss, W. (1972). *The domination of nature*. New York: Beacon Pub.

Levins, R., & Lewontin, R. (1985). *The dialectical biologist*. Cambridge: Harvard University Press.

Lewontin, R., Rose, S., & Kamin, L. (1984). *"Not in our genes" biology, ideology and human nature*. New York: Pantheon Books.

MacMahon, B. (1975). *Principios y métodos de la epidemiología*. México: La Prensa Médica Mexicana.

Marmot, M. G., & Wilkinson, R. G. (Eds.). (2006). *Social determinants of health* (2nd ed.). New York: Oxford University Press.

Marx, K. (1971). *Elementos fundamentales para la crítica de la economía política (Grundisse)* (pp. 1857–1858). México: Siglo XXI Editores.

Marx, K. (1972). *Elementos fundamentales para la crítica de la economía política (borrador) 1857–1858*. Buenos Aires: Siglo XXI.

Osorio, A. (1988). *Hamiltonian systems: Chaos and quantization*. Cambridge: Cambridge University Press, 238 p.

Otero, G., Pechlaner, G., Liberman, G., & Gürcan, E. (2015). The neoliberal diet and inequality in the United States. *Social Science & Medicine, 142*, 47–55. https://doi.org/10.1016/j.socscimed.2015.08.005.

Paredes, R. (1938). *El imperialismo en el Ecuador. Oro y sangre en Portovelo*. Quito - Ecuador: Editorial Artes Gráficas.

Pradeepa, R., & V Mohan, V. (2017). Prevalence of type 2 diabetes and its complications in India and economic costs to the nation European. *Journal of Clinical Nutrition, 71*, 816–824.

Punch, K. (2014). *Introduction to social research: Quantitative and qualitative approaches*. Los Angeles: Sage.

Ribeiro, S. (2016). *Cuarta revolución industrial, tecnologías en impactos*. El ciervo Herido, Blog de Omar González. https://elciervoherido.wordpress.com/2016/11/21/cuarta-revolucion-industrial-tecnologias-e-impactos-silvia-ribeiro/

Rosen, G. (1958). *A history of public health*. New York: MD Publications.

Samaja, J. (1993). *Epistemología y metodología: elementos para una teoría de la investigación científica*. Buenos Aires: Eudeba.

Samaja, J. (1996). *Epistemología y metodología*. Buenos Aires: Eudeba.

Santos, M. (1996). *A natureza do espaço: técnica e tempo; razão e emoção*. São Paulo: Editora Hucitec.

Sigerist, H. (1945). *Civilization and disease*. Ithaca: Cornell University Press.

Subirats, J. (2019, febrero). ¿Del poscapitalismo al postrabajo? *Nueva Sociedad, 279*, 34–48.

Tashakkori, A. T. C. (1998). *Combining qualitative and quantitative approaches (Applied social research methods series)* (Vol. 46). London: Sage.

Veraza, J. (2008). *Subsunción real del consumo al capital*. México: Editorial Itaca.

Virchow, R. (1848). Report on the Typhus Epidemic in Upper Silesia. *Medical Reform, 8*.

Waitzkin, H. (2011). *Medicine and public health at the end of empire* (pp. 9–10). Boulder, CO: Paradigm Publishers.

Waitzkin, H., Iriart, C., Estrada, A., & Lamadrid, S. (2001, October). Social medicine then and now: Lessons from Latin America. *American Journal of Public Health, 91*(10), 1592–1601. https://doi.org/10.2105/AJPH.91.10.1592

Waitzkin, H., & Working Group for Health Beyond Capitalism (Eds.). (2018). *Health care under the knife: Moving beyond capitalism for our health*. New York: Monthly Review Press.

Lost in Causality: How Epidemiology's Counterfactual Causal Inference Revolution Upholds Class, Race and Gender Inequities

Carles Muntaner and James R. Dunn

A fact in itself is nothing. It is valuable only for the idea attached to it...

Claude Bernard

Author of *Introduction à l'étude de la médecine expérimentale (1865)*

Abstract We critique empiricism, the dominant epistemology in epidemiology and public health from a scientific realist perspective. Building on our previous work, we also take on the popular counterfactual/potential outcomes epistemology based on its neglect of ontology and shunning causal mechanisms which are reduced to statistical methods (e.g., mediation). We the argue that ontology, epistemology, axiology and ethics constitute a philosophical system in epidemiology, and, in particular in social epidemiology and health equity/social inequalities research that ends up supporting capitalism, patriarchal/gendered and racialize social systems.

Keywords Scientific realism · Empiricism · Counterfactual · Potential outcomes · Ethics · Epidemiology · Public health · Social determinants of health · Social inequalities in health · Social epidemiology

In this chapter we draw attention to what we consider several features of the contemporary dominant epistemology in Epidemiology and Public Health (henceforth E&PH) that sustain our dominant social system, what we call Racialized Patriarchal Capitalism (RPC). In particular we argue that a mutually reinforcing relationship between Epistemology and Axiology (Values) in contemporary (E&PH) explains their support for RPC.

C. Muntaner (✉)
University of Toronto, Toronto, ON, Canada
e-mail: carles.muntaner@utoronto.ca

J. R. Dunn
McMaster University, Hamilton, ON, Canada

J. Vallverdú et al. (eds.), *Philosophical and Methodological Debates in Public Health*, https://doi.org/10.1007/978-3-030-28626-2_4

More specifically, the growing interest in causation in epidemiology in the last two decades (Kaufman and Poole 2000; Glass et al. 2013; Gage et al. 2016; Krieger and Davey Smith 2016) has led to a recent and vibrant debate around the counterfactual/potential outcomes approach to causal inference (Vandenbroucke et al. 2016; Glass et al. 2013; Chiolero 2016; Vanderweele et al. 2016; Schooling et al. 2016; Broadbent et al. 2016; Hernán 2016; Kaufman 2016; Schwartz et al. 2016; Krieger and Davey Smith 2016). Without questioning the necessity of causal inference investigations, we argue that debates about potential outcomes/counterfactual approaches are missing aspects of causality that are necessary to understand the type of evidence and interventions produced by E&PH and their political implications in upholding our social system (RPC). Drawing from debates on causation in social epidemiology (Muntaner et al. 1996, 1997; Cooper and Kaufman 1999; Muntaner 1999, 2013; Kaufman and Cooper 1999; Krieger and Smith 2000; Dunn 2012; Galea and Link 2013a; Galea and Link 2013b) we critique recent empiricist (i.e., counterfactual/potential outcomes) (Glass et al. 2013) and constructivist/empiricist (Krieger and Davey Smith 2016) epistemologies in social epidemiology in favor of a realist approach that encompasses both ontology, epistemology and axiology (Muntaner 2013). Although we select social epidemiology as the study area to illustrate our arguments, we anticipate that our conclusions might be generalizable to some additional epidemiologic subdisciplines.

Counterfactual/Potential Outcomes Empiricism Versus Scientific Realism: Causal Inference Versus Causation (and Causal Inference)

The main features of scientific realism stem from its ontology and epistemology. Realist ontology can be summarized by the statement "the world exists independently of our knowledge of it" (Sayer 1992) for our purposes, a population and the social phenomena we are examining exist independently of the epidemiologist who observes it. Ontological realism also claims that our senses are too limited to understand the world and that we need theories, which refer to the causal powers and liabilities of objects and the way that they interact, not only experience (i.e., data), to make sense of it. The scientific realist epistemology states that at least parts of the world can be known with the scientific method. Scientific realism (Smart 1963) is preferred to critical realism (Archer et al. 2007) because the later does not emphasize the scientific method as means to know about the world (Muntaner 2013).

The philosophical antecedents of the counterfactual/potential outcomes (CF/PO) popularized by Pearl (2018) (or the modern epidemiology of Rothman earlier) are British empiricists (Hume and the empiricist cluster of epistemologies that he inspired). In addition to his empiricist epistemology Hume was a defender of social inequality as per his elitism and attitudes about on race, indigenous peoples and colonialism (Prescott 2018). Such observation might be seen as an opportunistic attack on the epistemologist, but it is our claim that ontology and epistemology set constraint on values (Muntaner 1999, 2013). On the other hand, state of the art E&PH neglects the realist approach to medicine starting with Claude Bernard and more contemporary French scientists such as the biologist Francois Jacob.

The CF/PO approach does indeed centers around causal inference, and the empiricist epistemology has a well know reluctance to talk about causation (but not about causal inference). Note that even the epidemiologic critics of the CF/PO approach are reluctant to talk about causation preferring the language of "lack of theory" (Krieger 2000) or "inference to the best explanation" (Krieger and Davey Smith 2016). The problem with Lipton's "inference to the best explanation" is that it is empiricist which cannot lead to the trans-empirical constructs (i.e., meta-physics) which are needed in any advanced scientific explanation (Muntaner 1999). Thus the crucial influence of empiricism in current epidemiology is its neglect of ontology, causation as an unobserved real process. Pearl's causality "revolution" in epidemiology and AI is de facto only a renaissance of causal inference in epistemology. Pearl's causality starts late in the scientific process, when we need to decide whether, say, the virus causes the syndrome. Pearl's revolution is totally silent on the previous ontological step where theories are built with trans-empirical concepts and mechanisms inaccessible to direct observation with our senses: how the HIV virus via a biological mechanism generates the AIDS clinical syndrome.

The empiricist traditions ignore ontology because most causal mechanisms cannot be directly observed, and thus does not distinguish between indicators and their referents—the referents being the causal processes in reality. For example, the socio-psychological causal mechanisms underlying "residential racial segregation" or "institutional racism" (Muntaner et al. 1996) which we have an imperfect ability to observe (i.e., via indicators such as the "number of mortgages denied to Blacks in a given area" or "wage inequality between Blacks and Whites in a firm"). Thus in empiricist epidemiology (e.g., "racial disparities") the indicator (race) and the referent (racial discrimination) are collapsed and the account of the analysis is merely about "variables" and not the (real) individuals, structures and causal mechanisms that they represent. Indeed, the term 'indicator' is telling—it is an indicator *of something* and a large part of causal analysis should be identifying what that something it and how it behaves in the real world.

For example, economic inequality is the referent of indicators such as "income" or "occupation" and the gradient in health is reduced to the mere association between social indicator/variables and health indicator/variables, devoid of any reference to causal mechanisms (e.g., Marmot and Bell 2016). It is important to realize that in the Whitehall studies there is no mechanism linking "occupation" to "autonomy", "job control", "participation" or other psychosocial mediators, because occupation refers to the technical attributes of work and says little about social relations at work (Muntaner and O'Campo 1993; Muntaner et al. 2012). Marxian theoretical frameworks (Navarro 1982, 1991) or the contradictory class location (CCL) hypothesis include mechanisms linking class relations of domination and exploitation at work to psychosocial exposures such as lack of control or high job demands (Muntaner et al. 2015). In fact, the CCL hypothesis has been tested several times refuting the gradient

(Muntaner et al. 2015). If proponents of the gradient where consequent empiricists, they would consider the gradient hypothesis rejected with a single "Black Swan". That such simple association has been considered sufficient to epidemiologists for so many decades is remarkable and suggests, whether investigators are aware of it or not, that there is little interest in explaining (understanding via causal mechanisms) health inequalities.

An Epistemological Revolution or a Statistical Multiverse: The Counterfactual Approach to Causation

From all of the above we conclude that the epistemology of contemporary epidemiology remains empiricism, which states that scientific knowledge stems from experience ("observations" in epidemiologic terminology). A major flaw of empiricist epistemology, including the dominant potential outcomes framework (Glass et al. 2013), is to conflate epistemology (what can be known) from what it is (ontology, in reference to the world outside the observer). This narrow focus on the criteria for proof of a causal relation has limits our understanding of social systems. A narrow focus on the causal relation [a relation between events (Blackburn 2005), not between things or properties as it is often stated in contemporary epidemiology textbooks (Rothman et al. 2008)] neglects the role of uncovering non observable causal mechanisms. Absence of explanatory causal mechanisms limits the depth of what can be known (i.e., how things work), reducing epistemology to observable associations. The counter-factual revolution in the epistemology of epidemiology does not offer a solution to the lack of ontology (Pearl 2018). For example, its reliance on Bayesian statistics for causal inference (Pearl 2018) constitutes important epistemological and ethical problems. Radical Bayesian approaches to causal inference focus on subjective probabilities: they reduce epistemology, and therefore causal inference, to individual subjectivity (psychology). The ethical consequence is moral relativism, which deters public health action since there is no way to adjudicate between different subjectivities. Other versions of Bayesian approaches to causal inference are mere distractions from the quest in finding explanatory causal mechanisms (Howson and Urbach 1993; Pearl 2018). This presents another problem of scientific ethics since scientists should not engage in activities that deviate from the search for truth.

Beyond these problems of empiricism, there are other reasons why the counterfactual/potential outcomes approach is ill-equipped to adequately inform causation and causal inference. This stems from the notion that counterfactual sentences are not propositions whose truth can be determined. For example "If Michael had studied methods in genetic epidemiology, his impact factors would have been even higher" cannot be a hypothesis since "If Michael had studied methods in genetic epidemiology" is not a proposition, thus it cannot be tested for its factual truth. In other words, counterfactual sentences are not propositions whose truth content can be determined empirically. Therefore, counterfactuals have no significance to causal inference. We conclude that counterfactual comparisons, a foundation of observational epidemiology is a sheer imaginary, yet they are commonly taken as necessary hypotheses in causal inference.

As an illustration of this point, let's consider how the current use of counterfactuals in the epidemiology of "racial disparities" (Magnani et al. 2016). Counterfactual approaches to "race" take the direct observation of skin color or its self-report as observations from which causal inferences are made thus impeding explanations via causal mechanisms involving non observable trans empirical constructs concepts such as, for example, institutional racism (Muntaner et al. 1996; Muntaner 1999, 2013). This construct refers to mechanistic hypotheses in a social structure (that is, the sum of all social relations in a society). For example, we may hypothesise that institutional racism reduces access to labor markets, leading to chronic unemployment, lack of psychological control and high rates depression. Thus, testing hypotheses involving causal mechanisms would require data gathering with indicators such as hiring practices, promotion and wage differentials in firms, and individual behaviors and attitudes at the workplace. The empiricist counterfactual approach would not uncover any of those mechanisms, as it would be satisfied with inferences obtained from associations between "race" (Black vs. White categories) and health (e.g., Pearl 2018). It is not surprising that Pearl's Book of Why contains a defence of an author who advanced an essentialist/organismic concept of race (Muntaner et al. 1996). The price for conflating causality with the proof for causality (criteria for causal inference) as seen in the counterfactual/potential outcome approach is the lack of explanations with mechanisms.

The use of the race variable is only one example where the empiricist conflation of ontology with epistemology in social epidemiology leads to limited explanations via causal mechanisms and ultimately contributes to maintain the social order (i.e. racial, class and gender inequalities in health). Sayer's distinction between "indicator" and "referent" is important here (Sayer 1999). The social sciences and psychology contain substantial theoretical work on the relationship between a causal process (referent) and the way it is measured (indicator) (Sayer 1999). The ubiquitous gradient (Marmot and Bell 2016) referring to associations between income (or occupation) and health illustrates the consequences of this absence in epidemiology. There is no social relational mechanism that might explain the occupational class gradient in health (Marmot and Bell 2016).

We can obtain data on income and observe and association with health, but we do not explain either the social mechanisms leading to income differentials, nor the mechanisms linking income to poor health in doing so. There is no mechanism in gradient studies that could explain how occupation or income strata might be differently associated with health indicators. We can be confident that the health effects of income are not about eating it or rubbing it on our skin, so we can dismiss these causal mechanisms, but what are the mechanisms that matter? These tend to be lacking of a priori theories. What are the causal powers and liabilities of money? Under what conditions are these exercised? How might these factors lead to the repeated pattern of health differentiation by income level? We claim that all the evidence across time and place of associations between income gradients (e.g., Stringhini et al. 2017) and health could not shed light on the social mechanisms that might explain why say, a high income person might be healthier than a low income person. In fact, from an empiricist epistemology, once when a single explanation based on social and psychological mechanisms such as the contradictory class location is confirmed (e.g., Muntaner et al. 1998), the universally accepted generality of the gradient (Stringhini et al. 2017) should be refuted (see Muntaner et al. 2015 for a review). From an empiricist epistemology all it takes is a single "Black swan" to refute the law of the social gradient. Therefore lack of explanatory causal mechanisms such as class exploitation or domination will ensure that no political or policy intervention emerging from social epidemiology might be formulated to reduce class inequalities in health.

In light of the above, the call of empiricists, both proponents (Hernán 2016) and critics (Schwartz et al. 2016) of the CF/PO approach, for better defined exposures and interventions, is mistaken. What are really needed are not definitions but theoretical models that, upon empirical confirmation, reveal explanatory causal mechanisms on which to base political/policy interventions (Muntaner 2013). In other words if the experiment/clinical trial with well-defined interventions is the "gold standard", the experiment/clinical trial with explanatory mechanisms is the "rhodium standard".

An Appraisal of the Constructivist/Empiricist Alternative

In their critique of the CF/PO approach some researchers (Krieger and Davey Smith 2016) have proposed a causal inference alternative, the Inference to the Best Explanation (IBE). The author's approach is also empiricist, as their epistemology begins with observations/data but advocates for a plurality of methods of causal inference (including the CF/PO approach) and Inference to the Best Explanation (IBE) in particular. IBE's inference begins with observations and compares a set of competing hypotheses to decide which of them is best supported by the evidence.

Inductivism, the building block of IBE, implies that we "start with data". As a consequence it can never lead to deep explanations, for example those requiring theoretical constructs such as gravitational mass or racist ideology, because our senses are too limited (Muntaner 1999).

In fact CF/PO and IBE are incompatible since IBE begins with facts, objective occurrences, while CF relies on imaginary scenarios (see section above). Therefore, the authors could have rejected the CF/PO, rather than considering CF/PO one of their approaches to causal inference. Consistent with constructionism pluralism calls for consensus between contending views of causal inference and causation. There is also a veil of constructionism in the subjective nature of choosing among hypotheses (i.e. the "loveliest explanation"). IBE similarities with CF/PO are nonetheless substantial. There is no distinction between epistemology and ontology, observations are not informed by theory testing, and explanations are post-hoc. One big advantage of the IBE methods over the CF/PO approach is that IBE highlights the importance of mechanisms, a feature of advanced explanations, albeit post-hoc, that is, as heuristic or sound speculation. In sum the IBE cannot provide an alternative to the development of social epidemiology as it is burdened by many of the shortcomings of the CF/PO framework.

We also question the adequacy of the "pluralist" approach to causal inference and causation (Vandenbroucke et al. 2016). Primarily, pluralism in causal inference and methods is suggestive of constructivism (researchers construct their own proof for causation). Historically, pluralism is suggestive of Berkeley's subjectivist empiricism. Subjectivism renders science unnecessary since there is no objective truth. Truth is in the eye of the beholder, and "your truth is as good as mine."

The Causal Inference Bubble

The popularity of counterfactual models in causation seems a distraction from finding models with causal mechanisms that can be empirically tested (Muntaner 1999). Indeed a focus on counterfactuals does not require knowledge of the mechanisms at play (as in "Black and white disparities" or "SES gradients in health". Therefore, it ensures that its findings will not be relevant for policies or interventions, which require the precise cause-effect relationships that only mechanisms can provide. That is why it is politically innocuous to avoid theories and causal explanations via mechanisms and rely on associations (e.g., focus on differences between Black and White categories or groups of occupations). Mechanisms are necessary to understand in depth, and this is a major challenge, but even when the mechanisms of causation of a phenomenon are well understood, a practically adequate understanding of the mechanism of amelioration (or intervention) may be quite different. We can't just ram the causal mechanism into reverse as we would a car transmission and expect the phenomenon to be reversed. Lack of well-developed accounts of the causal mechanisms that produce health and an account of the mechanisms by which possible interventions to ameliorate inequalities in health might work is what is lacking in social epidemiology, not a lack of causal inference.

Social epidemiologists content with associations between Black/White comparisons and ranks of income/occupation and health indicators should consider that for social epidemiology to progress we need to generate unobservable constructs integrated in theoretical models (e.g., to uncover the social mechanisms that explain why a person with a supervisory/managerial occupation might present worse health than one with a worker occupation). Unobservable relations and mechanisms between social constructs (e.g., the 'social structure') help us make sense of the world because our experiences are too restricted. For example, although we cannot directly observe "nations" or "black holes" (constructs), we can observe them indirectly with population surveys and visible light, respectively. It is it not possible to advance causal models without moving away from an empiricist epistemology bounded by observations and repeated associations. The number of times something happens has nothing to do with why it happened, although sometimes repetition of patterning of events can point to the existence of causal processes that we still need to theorize. WWII only happened once, but we can still develop an account of the causal mechanisms that led to it.

Social and Public Health Policy Consequences

A positive heuristic of this debate has been the acknowledgment advanced by potential outcomes proponents of the link between causal inferences in epidemiology with public health programs (Glass et al. 2013). Yet without confirmed causal models it is difficult to design useful public health interventions or policies, since we may not know "what component of the intervention causes the desired effect". If the experiment/clinical trial is the empiricist's Golden Standard, we need to strive for the realist's Painite Standard with mechanisms.

The counterfactual/Potential Outcomes and other empiricist approaches are politically and policy conservative as they tend to favor the "status quo" (Muntaner 2013). If we do not know what causal mechanisms produce health inequities, how can we modify them? The recent history of social epidemiology is full of programs that generate vast amounts of data on associations ("the gradient", income inequality, race/ethnic disparities) but few causal mechanisms, therefore no interventions. The potential outcomes framework is used to determine what interventions are feasible on the basis of variables (e.g., race, income). Yet the lack of hypothetical causal social mechanisms limits both our knowledge and the possibility of interventions. Hernán (2016) recently noted that an intervention on "race" (Glymour and Glymour 2014) is not feasible because "race" is "ill defined". Yet the problem here is not one of definition of race (Hernán 2016) or "lack of well-defined exposures (Schwartz et al. 2016) or lack of theory (a hypotetico-deductive system, not a narrative convention (e.g. Krieger 2000), it's that the epidemiological literature on "race disparities" does not provide evidence of causal mechanisms to explain the associations between "race" and health. Testing theoretical causal mechanisms of interpersonal racism, institutional racism, labor market discrimination, or residential segregation could lead to successful interventions since we would know what cause (e.g., denying mortgages to dark skinned persons) produces what effect (e.g., dark skinned persons do not have access to health promoting residential neighborhoods). Using scientific constructs such as "racial discrimination" and "social causal mechanisms" integrated in theoretical models about how the world works is a previous step to confirming a causal mechanism. Once a causal mechanism is confirmed empirically with the help of indicators (number of mortgages denied to Latino families, socioeconomic indicators of Latino families' residential neighborhoods) we can apply this knowledge about causal mechanisms to design effective interventions. Therefore explanations via causal mechanisms (i.e., happening in the world independently of epidemiologists) are both distinct from proof of causation (e.g. Hill's criteria) and essential for effective interventions. Thus, interventions explained by causal mechanisms are more effective than those stemming by trial and error, "black box"/association generalizations, or personal experience.

On Saving FACE (Or Don't Shoot the Statistical Messenger)

In a recent article Daniel et al. (2016) (DDV) take on the recent critiques of the formal approach to quantitative causal inference (FACE; Vandenbroucke et al. 2016) and Krieger and Davey Smith (2016). DDV question the claim by these critiques that the FACE approach restricts the study of causal inference to humanly feasible interventions. The authors in our view correctly assert that FACE does not limit the type of interventions possible:

> We view the FACE as precisely offering formal tools to investigate cause–effect relationships. They are always guided by what KDS call IBE (inference to the best explanation). Indeed, IBE is often how one comes to investigate the specific cause–effect relationship in the first place.

The absence of theories with causal social mechanisms (an ontological problem) is what's missing in social epidemiology and public health, which is not a question of causal inference. The formal approach including DAGs are valuable tools to increase the exactness of theories and serve as a map for testing hypotheses and establishing causal inference with data analysis. The limitation of DAGs is not DAGs themselves, but the unwillingness of social epidemiologists to create, imagine causation as it occurs in the world—to theorize about the behavior of referents (real world phenomena) and not just variables (indicators). Before we determine causal inference (epistemology) we need to propose causation (ontology). Empiricists are reluctant to generate social causal mechanisms because they are skeptical of what cannot be directly observed (causal mechanisms involve constructs such as exploitation, sexism, racism that can only be inferred via indicators, not directly observed) therefore they prefer to stick to observations of skin color, sex, income, occupation, education and their associations. Absence of social causal mechanisms precludes causality upon which policies can intervene. Societies don't change. Empiricism thus leads to conservatism (or vice versa).[1] The implicit conservatism of the epistemology in social epidemiology reminds us of the embrace of another conservative epistemology in the 1980s and 1990s, namely Popperian falsacionism. Falsationism states that all we can know is that a proposition is not false. In that epistemology truth is elusive, therefore interventions postponed ad infinitum.

[1] A precursor of establishing the link between empiricism and capitalism is Lenin's 1909 "*Materialism and Empirio-Criticism. Critical Comments on a Reactionary Philosophy*".

Conclusion: Is There a Future for Causation in Social Epidemiology?

Because epidemiology is dominated by an empiricist epistemology that shuns explanation via causal mechanisms in favour of associations, the early adoption of methodological innovations in causal inference are considered most important, as in Pearl's "causation revolution" (Pearl 2018). In epidemiology and public health we still observe a strong adherence to British empiricism as in the incongruous cult of David Hume (Pearl 2018) where all we can know are observations. On the other hand, scientific realism, which separates ontology (causation as it exists in the world) and epistemology (causal inference), is greatly overlooked, as in the seminal work of Claude Bernard on internal medicine. Yet most developed sciences progress thought the understanding of systems, their components, relations, and their mechanisms (e.g. Mazzarello 1999; Kandel et al. 2000). This shortcoming seems particularly damaging to social epidemiology, which suffers from both a dearth of theories proper (hypothetic deductive systems) and explanatory social mechanisms.

References

Archer, M., Bhaskar, R., Collier, A., Lawson, T., & Norrie, A. (2007). *Critical realism: Essential readings*. London: Routledge.

Bernard, C. (1865). *Introduction à L'étude de la Médecine Expérimentale*. Paris: Baillière.

Blackburn, S. (2005). *The Oxford dictionary of philosophy*. Oxford: Oxford University Press.

Broadbent, A., Vandenbroucke, J., & Pearce, N. (2016). Author's reply to: VanderWeele et al., Chiolero, and Schooling et al. *International Journal of Epidemiology, 45*(6), 2203–2205. https://doi.org/10.1093/ije/dyw163.

Chiolero, A. (2016). Counterfactual and interventionist approach to cure risk factor epidemiology. *International Journal of Epidemiology, 45*(6), 2202–2203. https://doi.org/10.1093/ije/dyw159.

Cooper, R. S., & Kaufman, J. S. (1999). Is there an absence of theory in social epidemiology? The authors respond to Muntaner. *American Journal of Epidemiology, 150*(2), 127–128. https://doi.org/10.1093/oxfordjournals.aje.a009971.

Daniel, R. M., De Stavola, B. L., & Vansteelandt, S. (2016). Commentary: The formal approach to quantitative causal inference in epidemiology: Misguided or misrepresented? *International journal of epidemiology, 45*(6), 1817–1829.

Dunn, J. R. (2012). Explanation, philosophy and theory in health inequalities research: Towards a critical realist approach. In P. O'Campo & J. R. Dunn (Eds.), *Rethinking social epidemiology: Towards a science of change* (pp. 23–42). Dordrecht: Springer.

Gage, S. H., Hickman, M., & Zammit, S. (2016). Association between Cannabis and psychosis: Epidemiologic evidence. *Biological Psychiatry, 79*(7), 549–556.

Galea, S., & Link, B. G. (2013a). Six paths for the future of social epidemiology. *American Journal of Epidemiology, 178*(6), 843–849. https://doi.org/10.1093/aje/kwt148.

Galea, A., & Link, B. G. (2013b). Galea and link respond to 'pathologies of social epidemiology,' 'social epidemiology and scientific realism,' and 'off-roading with social epidemiology'. *American Journal of Epidemiology, 178*(6), 864. https://doi.org/10.1093/aje/kwt147.

Glass, T. A., Goodman, S. N., Hernán, M. A., & Samet, J. M. (2013). Causal inference in public health. *Annual Review of Public Health, 34*, 61–75. https://doi.org/10.1146/annurev-publhealth-031811-124606.

Glymour, C., & Glymour, M. R. (2014). Commentary: Race and sex are causes. *Epidemiology, 25*(4), 488–490.

Hernán, M. A. (2016). Does water kill? A call for less casual causal inferences. *Annals of Epidemiology, 26*(10), 674–680. https://doi.org/10.1016/j.annepidem.2016.08.016.

Howson, C., & Urbach, P. (1993). *Scientific reasoning: The Bayesian approach paperback.* Retrieved from https://www.amazon.ca/Scientific-Reasoning-Bayesian-Colin- Howson/dp/081269578X

Kandel, E. R., Schwartz, J. H., & Jessell, T. M. (Eds.). (2000). *Principles of neural science.* New York: McGraw-Hill.

Kaufman, J. S. (2016). There is no virtue in vagueness. *Annals of Epidemiology, 26*(10), 683–684. https://doi.org/10.1016/j.annepidem.2016.08.018.

Kaufman, J. S., & Cooper, R. S. (1999). Seeking causal explanations in social epidemiology. *American Journal of Epidemiology, 150*(2), 113–120. https://doi.org/10.1093/oxfordjournals.aje.a009969.

Kaufman, J. S., & Poole, C. (2000). Looking back on 'causal thinking in the health sciences. *Annual Review of Public Health, 21*(1), 101–119. https://doi.org/10.1146/annurev.publhealth.21.1.101.

Krieger, N. (2000). Epidemiology and social sciences: Towards a critical reengagement in the 21st century. *Epidemiologic Reviews, 22*(1), 2000–2155.

Krieger, N., & Smith, G. D. (2000). Re:"Seeking causal explanations in social epidemiology". *American Journal of Epidemiology, 151*(8), 831–832.

Krieger, N., & Davey Smith, G. (2016). The tale wagged by the DAG: Broadening the scope of causal inference and explanation for epidemiology. *International Journal of Epidemiology, 45*(6), 1787–1808. https://doi.org/10.1093/ije/dyw114.

Magnani, J. W., Norby, F. L., Agarwal, S. K., Soliman, E. Z., Chen, L. Y., Loehr, L. R., & Alonso, A. (2016). Racial differences in atrial fibrillation-related cardiovascular disease and mortality: The atherosclerosis risk in communities (ARIC) study. *JAMA Cardiology, 1*(4), 433–441. https://doi.org/10.1001/jamacardio.2016.1025.

Marmot, M., & Bell, R. (2016). Social inequalities in health: A proper concern of epidemiology. *Annals of Epidemiology, 26*(4), 238–240.

Mazzarello, P. (1999). A unifying concept: The history of cell theory. *Nature Cell Biology, 1*(1), E13–E15.

Muntaner, C. (1999). Social mechanisms, race, and social epidemiology. *American Journal of Epidemiology, 150*(2), 121–126.

Muntaner, C. (2013). Invited commentary: On the future of social epidemiology—A case for scientific realism. *American Journal of Epidemiology, 178*(6), 852–857. https://doi.org/10.1093/aje/kwt143.

Muntaner, C., & O'Campo, P. J. (1993). A critical appraisal of the demand/control model of the psychosocial work environment: Epistemological, social, behavioural and class considerations. *Social Science and Medicine, 36*(11), 1509–1517.

Muntaner, C., Nieto, F. J., & O'Campo, P. (1996). The bell curve: On race, social class, and epidemiological research. *American Journal of Epidemiology, 1444*(6), 531–536.

Muntaner, C., Nieto, F. J., & O'Campo, P. (1997). Additional clarification re: 'The bell curve: On race, social class, and epidemiological research'. *American Journal of Epidemiology, 146*(7), 607–608.

Muntaner, C., Edwin, N. G., & Chung, H. (2012). *Better health: An analysis of public policy and programming focusing on the determinants of health and health outcomes that are effective in achieving the healthiest populations,* Canadian health service research foundation. Retrieved from https://www.cfhi-fcass.ca/sf-docs/default- source/commissioned-research-reports/Muntaner-BetterCare-EN.pdf?sfvrsn=0

Muntaner, C., Edwin, N. G., Chung, H., & Prins, S. J. (2015). Two decades of neo- Marxist class analysis and health inequalities: A critical reconstruction. *Social Theory and Health, 13*, 267–287.

Muntaner, C., Eaton, W. W., Diala, C., Kessler, R. C., & Sorlie, P. D. (1998). Social class, assets, organizational control and the prevalence of common groups of psychiatric disorders. *Social Science and Medicine, 47*(12), 20143–22053.

Navarro, V. (1982). The labor process and health: A historical materialist interpretation. *International journal of Health Services, 12*(1), 5–29.

Navarro, V. (1991). Production and the welfare state: The political context of reforms. *International Journal of Health Services, 21*(4), 585–614.

Pearl, J. (2018). Comments on: The tale wagged by the dag. *International Journal of Epidemiology, 47*(3), 1002–1004.

Prescott, W. H. (2018). *Revival: History of the conquest of Mexico (1886) with a preliminary view of the ancient Mexican civilization and the life of the conqueror, Hernando Cortes* (2nd ed.). London: Routledge.

Rothman, K. J., Greenland, S., & Lash, T. L. (2008). *Modern epidemiology*. Retrieved from https://www.amazon.ca/Modern-Epidemiology-Kenneth-Rothman-2008-03- 14/dp/B01FKTHY18

Sayer, A. (1992). *Methods in social science: A realistic approach, Methods in social science* (2nd ed.). London: Routledge.

Sayer, A. (1999). Bourdieu, Smith and disinterested judgement. *The Sociological Review Foundation, 47*(3), 1–13.

Schooling, C. M., Chow, C. K., & Young, S. A. (2016). Causality and causal inference in epidemiology: We need also to address causes of effects. *International Journal of Epidemiology, 45*(6), 2200–2201. https://doi.org/10.1093/ije/dyw160.

Schwartz, S., Gatto, N. M., & Campbell, U. B. (2016). Causal identification: A charge of epidemiology in danger of marginalization. *Annals of Epidemiology, 26*(10), 669–673. https://doi.org/10.1016/j.annepidem.2016.03.013.

Smart, J. J. (1963). *Philosophy and scientific realism*. Atlantic Highlands, NJ: Humanities Press.

Stringhini, S., Carmeli, C., Jokela, M., Avendaño, M., Muennig, P., Guida, F., Ricceri, F., d'Errico, A., Barros, H., Bochud, M., Chadeau-Hyam, M., Clavel-Chapelon, F., Costa, G., Delpierre, C., Fraga, S., Goldberg, M., Giles, G. G., Krogh, V., Kelly-Irving, M., Layte, R., Lasserre, A. M., Marmot, M. G., Preisig, M., Shipley, M. J., Vollenweider, P., Zins, M., Kawachi, I., Steptoe, A., Mackenbach, J. P., Vineis, P., Kivimäki, M., & LIFEPATH Consortium. (2017). Socioeconomic status and the 25 × 25 risk factors as determinants of premature mortality: A multicohort study and meta-analysis of 1·7 million men and women. *Lancet, 389*(10075), 1229–1237.

Vandenbroucke, J. P., Broadbent, A., & Pearce, N. (2016). Causality and causal inference in epidemiology: The need for a pluralistic approach. *International Journal of Epidemiology, 45*(6), 1776–1786.

VanderWeele, T. J., Hernán, M. A., Tchetgen, E. J., & Robins, J. M. (2016). Re: Causality and causal inference in epidemiology: The need for a pluralistic approach. *International Journal of Epidemiology, 45*(6), 2199–2200. https://doi.org/10.1093/ije/dyw162.

Coeliac Disease and Non-coeliac Gluten Sensitivity: A Clinical and Philosophical Discussion

F. Borrell-Carrió, E. Peguero-Rodríguez, V. Morales-Hidalgo, J. Castellote-Alonso, and A. Estany

Abstract The questioning of the introduction of new diseases is a typical feature of psychiatry: Does ADHD (attention deficit/hyperactivity disorder) exist? Why "hysteria" has not disappeared from diagnosis taxonomies? Here, we show that this characteristic is not exclusive to psychiatry. In the case of coeliac disease and non-coeliac gluten sensitivity (NCGS), an important tension arises between a pragmatist position, in favour of diagnosing NCGS, and an opposing essentialist conception, since no "biological core" specific to NCGS can be identified. When proposals for new diseases appear, practitioners adopt positions in favour or against their acceptance, supported by different epistemological approaches. In this work, we analyse these approaches, which lie behind many nosological and taxonomic discussions in medicine.

Keywords Coeliac disease · Non-coeliac gluten sensitivity · Disease script · Pragmatism · Essentialism · Invariantism · Contextualism

---

F. Borrell-Carrió (✉)
Department of Clinical Sciences, University of Barcelona, Barcelona, Spain

Gavarra EAP (Primary Healthcare Team), ICS (Catalan Regional Health Service), Barcelona, Spain
e-mail: 12902fbc@comb.cat

E. Peguero-Rodríguez
Department of Clinical Sciences, University of Barcelona, Barcelona, Spain

Castelldefels EAP (Primary Healthcare Team), ICS (Catalan Regional Health Service), Barcelona, Spain

V. Morales-Hidalgo
Garraf-Penedés Paediatric Unit, ICS (Catalan Regional Health Service), Barcelona, Spain

J. Castellote-Alonso
Bellvitge Teaching Hospital, Barcelona, Spain

A. Estany
Department of Philosophy, Autonomous University of Barcelona, Barcelona, Spain

J. Vallverdú et al. (eds.), *Philosophical and Methodological Debates in Public Health*, https://doi.org/10.1007/978-3-030-28626-2_5

Introduction

Do diseases have a real existence or are they human constructions? Are they in the minds of doctors or do they constitute a physical and biological reality?

Nobody would doubt that a tumour has a real presence; but, what about schizophrenia? Just as in the case of a tumour, we find the same situation with coeliac disease (CD). It has a set of unquestionable biological markers, and although the range of its clinical manifestations sometimes leads to late (and even erroneous) diagnoses, the genetic, histological, biochemical and physiological bases of CD mean that it is a "perfect" model of a disease. In some sense, we could also talk of it being a deterministic model, as the presence of certain characteristics or specific results predict with certainty a set of further facts, either at that moment or at a later time. It is precisely a predictable set of findings that allows us to formulate a prognosis that we call "coeliac disease".

The question that interests us here though is whether even such a perfect disease as CD could become blurred around the edges and might start to blend in or meld with other less well-defined conditions; "less real", we might say, to put it in different terms.

In effect, as Diaz and co-workers claim, examinations of some patients have provided an increasing body of "*evidence suggesting the existence of a new entity: non-coeliac gluten sensitivity (NCGS). For many years, these patients have been misdiagnosed with irritable intestine syndrome (IIS), depression or fibromyalgia, while being kept on a diet that includes gluten and, in some cases, being referred for psychiatric evaluation*" (Díaz Marugán et al. 2013).

The aim of the present work is to show how and why the introduction of a new entity, in this case NCGS, initially provokes a sceptical response from doctors, and later leads them to understand CD in the broader context of multiple food intolerances.

NCGS: Problem Resolution or Complication?

NCGS presents clinically as very similar to CD, but on a lesser scale, with episodes of diarrhoea which resemble irritable intestine syndrome (IIS), but also, in order from most to least common: "*a lack of concentration, tiredness, eczema and skin rashes, headaches, arthralgia and myalgia, muscle cramps, depression and anaemia*" (Díaz Marugán et al. 2013). Together, CD and NCGS could affect as much as 10% of the general population, with an average age at onset of 40 (ranging from 17 to 63), that is, it would no longer be considered a predominantly paediatric illness.

We can only reliably demonstrate NCGS in patients with "ad hoc" symptoms that are the result of eating gluten, that do not present an allergy to other proteins in wheat and which do not present intestinal alterations in a biopsy (Catassi et al. 2013). This is where an initial problem immediately arises: once a doctor suggests a gluten-free diet and the patient experiences an improvement, how can we persuade that patient to start eating gluten again, so that we can perform an intestinal biopsy? Therefore, most clinics resort to a trial and error method: include or exclude gluten from the diet and see what happens (Volta et al. 2013). Then, if the patient improves on a gluten-free diet and has no other markers of CD, diagnose NCGS.

This new entity appeared as a refuge for patients who previously had received no satisfactory or plausible diagnosis. Perhaps they had been classified as having IIS or a food intolerance, or as suffering from anxiety. However, at the same time as the new diagnosis offered relief to some, it meant that cracks were starting to form in what had been an almost perfect model of disease: CD. Therefore, we should not be surprised at all that at first there was a general reluctance to recognise NCGS: could we not be witnessing a pseudo-disease, some new effect of the powerful human imagination or simply of the growing "anti-gluten" fashion? Only the publication of a series of clinical cases in different countries cleared the way for the current acceptance of NCGS. So, should we see this as a triumph of pragmatism over essentialism?

The case is not quite so straightforward and we must not jump the gun here. Many experts believe that this novel entity, NCGS, may undergo important changes in the near future. On one hand, these may be in relation to gluten as the agent that is chiefly responsible for the disorder, and also in relation to future biological markers:

> many other factors also contribute, including low-grade intestinal inflammation, increased intestinal barrier function and changes in the intestinal microbiota. Gluten and other wheat proteins, such as amylase trypsin inhibitors, are the primary triggers of this syndrome, but it has also been hypothesized that a diet rich in fermentable monosaccharides and polyols may elicit its functional gastrointestinal symptoms (Volta et al. 2013).

On the other hand, though along similar lines, Biesiekierski et al. propose a broadening of the concept of NCGS to that of a reaction to FODMAPs (fermentable, oligo-, di-, monosaccharides and polyols). This would cover patients who up until now have been classified as having IIS and yet who benefit from a reduction of these wheat derivatives (and not just gluten) (Biesiekierski et al. 2011).

This double perspective frames efforts to find biological markers associated with intolerance to these molecules, or changes in the intestinal biota, which would provide a more solid base on which to establish an aetiological diagnosis. Therefore, it would seem that for the patient, it might at least fulfil the role of answering the question: "Doctor, what's wrong with me?" But that possible answer would come at the price of a whole new set of tasks to be completed: bringing coherence to this new diagnosis, and discovering its "anchors" that ground it within the taxonomy of illnesses (we will return to this concept of "anchor" in section "Have We Left Ourselves Without a Model of Disease?"). So, rather than having been defeated, it is actually the case that the essentialist camp is working hard to find a more general model and reframe NCGS as a "specific case" of a whole series of intolerances to wheat proteins. We must not forget that the objective of every good essentialist doctor is fit together all the different human illnesses in a single model that is as simple as possible, and which provides strong biological anchors with predictive power (or, in medical terms, with prognostic capacity).

Pragmatism, Perplexity and a Model of Perfection

Before moving on any further, we first wish to clarify some of the key concepts involved here; the first of these is the notion of a model of perfection. We human beings, immersed in our different cultures, take on board different models of perfection from the time we are children; models that indicate to us how people and things work, and how we should behave. The norms for being polite in our social interactions, for example, obey a model of perfection (Brown and Levinson 1987). Similarly, in Health Sciences, there is an ideal model of disease: a known noxa produces a physiological disorder which we remedy through the application of a specific treatment (Borrell and Estany 2016). When we find no explanation for a patient's suffering through the use of a "classic"—or maybe we should say *orthodox*—model, a state of perplexity arises and it becomes impossible to act, as we normally would, motivated by our belief (Peirce 1877). This would be the case of a doctor who suspects that a patient is suffering from CD, the gluten-free diet improves the patient's symptoms, but the test for the specific CD antibodies is negative, as is the intestinal biopsy. In such a case, the doctor is faced with the dilemma of whether or not to pursue the study of this specific clinical case further; whether the additional effort will be worthwhile.

The philosophical tradition of pragmatism postulates that there is only a difference between two theories when they lead to different practical results (James 1997). The tasks of the scientist should therefore revolve around finding the differences between two theories, or models of disease, that lead to different results (for example, in terms of morbidity and mortality). For this reason, professions who work in medicine are necessarily pragmatists, as they do not dedicate time and effort to *situations of avoidable perplexity* (Table 1). And this is where a series of ethical questions arise, such as, for example, whether a doctor should investigate whether a 95-year-old patient with a liver tumour and suffering from cachexia has a hepatocellular carcinoma or a metastatic tumour. This is certainly a diagnostic effort that the majority of doctors would avoid, since it is not necessary to offer the patient the palliative care needed. Knowing whether we are faced with a primary cancer or metastasis, in this case, no longer represents a therapeutic advantage. If we move

Table 1 Types of clinical perplexity

Perplexity—What, up until now, was considered to be the "best solution" or the "best course of action" no longer seems to qualify as such, and consequently we do not know how to act
Unavoidable perplexity—The clinical problem is not a standard or common one and it requires us to choose between several possibilities. To do so, we need to reconsider the beliefs on which we construct arguments in favour of and against each possible course of action. *The need to choose stems from the fact that we perceive an even greater risk if we do not make a choice*
Avoidable perplexity—Situation in which we can avoid the state of perplexity in a beneficial way, either because it is not completely necessary to make a decision, or because the choice between the possibilities—Whatever they may be—Will not change the unfolding of events. Therefore, the course of action chosen—even choosing not to do anything—does not increase the risk of an avoidable adverse event

back to the case of a patient whose symptoms improve after following a gluten-free diet, here we must similarly question whether it is worthwhile attempting to determine whether the underlying cause is CD or NCGS (or even some other rarer disease) and subject the patient to an invasive intestinal biopsy. This brings us to the analysis of the "illness script" or "disease script".

The Illness Scripts *for Coeliac Disease and Non-coeliac Gluten Sensitivity*

To examine in more depth how doctors understand diseases, a diagnostic model has been put forward that consists of *illness scripts* or *disease scripts* (Schmidt and Rikers 2007; Van Schaik et al. 2005). These scripts contain the different ways in which illnesses present themselves and evolve, and distinguish between three bodies of knowledge that clinicians acquire theoretically but especially (and very importantly) through empirical observation: (a) how the disease presents (encapsulation), (b) the necessary diagnostic criteria and (c) disease evolution. Illness scripts constitute a type of knowledge which is in part theoretical and in part tacit. For example, maybe we could give a lecture on CD, but faced with an individual patient, perhaps we would say "this doesn't seem to fit with CD". In such a case, a type of tacit knowledge (Schön 1983) comes to mind, which we would not typically be able to explain if we were not faced with a specific clinical situation and so it is not something that we would have included in the lecture we gave.

Let us now consider the illness script of CD compared to that of NCGS:

(a) Examination of the clinical presentation (also called knowledge encapsulation). We consider the possibility of CD when we are faced with a patient (especially paediatric) with chronic episodes of diarrhoea that affect the patient's nutritional status, but also in adults with diarrhoea that we would often characterise as and label as IIS, dermatological patients, (pruritus, with or without compatible lesions), haematological patients, (unexplained anaemia) and more rarely psychiatric problems. NCGS would not present any different features except less severity of all these presentations; *therefore there is no change in the illness script as far as disease presentation is concerned.*

(b) Diagnostic criteria. In theory, both CD as NCGS require an intestinal biopsy from the symptomatic patient. This requirement has been relaxed in the case of CD, as just four of the five criteria given by Cattassi (Catassi and Fasano 2010) are now preferred as a requirement:

> The biopsy, considered the gold standard of diagnosis, has recently been questioned as a conclusive test. In fact, the wide variety of findings related to coeliac disease suggests that it is difficult to conceptualise the diagnosis process in rigid algorithms, which do not always cover the clinical complexity of this disease. In contrast, we find a quantitative approach that can be defined as the "4 of 5" rule to be clinically useful: the diagnosis of coeliac disease is confirmed if at least 4 of the following 5 criteria are met: (1) typical symptoms of coeliac disease; (2) high positive titres of IgA antibodies; (3) haplotype HLA-DQ2 or DQ8; (4) intestinal biopsy; and (5) response to a gluten-free diet.

In the case of NCGS, is it enough that the clinical situation improves thanks to a diet free of gluten to avoid this biopsy? Here, there is no clear consensus; but in practice, this is indeed what tends to happen (Catassi et al. 2015). Therefore, in both CD and NCGS the requirement of an intestinal biopsy has been relaxed, leaving it to the other four factors to determine whether we diagnose CD or NCGS. Otherwise, we would be in a situation of avoidable perplexity: it is not worth studying the patient further if we will arrive at the same conclusion that we should prescribe a gluten-free diet.

(c) Variation of the therapeutic response and evolution of the prognosis. Here, the most fundamental aspect is that once the intake of gluten was eliminated, the patient improved, and reintroducing gluten into the diet produces renewed suffering. Also, if after a certain time the patient develops symptoms again, then the diagnosis has to be reconsidered. It is true that we still do not know many of the details of NCGS, for example: whether it is plausible to suspect NCGS when faced with patients with no intestinal symptoms; whether NCGS can develop into CD; or whether the two conditions represent different locations along a single continuum. However, these aspects do not seem to be relevant here and we can conclude that there is no difference between the *script* for CD and that for NCGS in this case either.

So, the disease script for NCGS is the same as that for CD in everything except one fundamental aspect: *we will continue to suspect that NCGS is the correct diagnosis when the test for specific CD antibodies is negative.*

Perplexity and Bayes' Theorem

Based on this illness script we can imagine the following situation:

> A patient presents with diarrhoea and tests negative for antibodies; the patient's situation improves on a gluten-free diet. We now have two different possibilities:
>
> 1. Currently, many clinics would be content with a diagnosis of NCGS and would not proceed with any further studies, unless other diagnoses were suspected.
> 2. Before NCGS was recognised, a proportion of these patients would probably have received a diagnosis of IIS and others would have been recommended for a colonoscopy or biopsy.

Possibility 1 is a scenario of *avoidable perplexity* (Table 1): the doctor believes that it is not worthwhile investigating further since, in any case, the same conclusion will be reached: prescribe a gluten-free diet.

Possibility 2, in contrast (prior to the recognition of NCGS as an ontological and nosological entity), is less satisfactory. It has always been recognised that IIS is a catch-all "last resort" diagnosis, which we only arrive at after reliably ruling out other pathologies. The sensation of *unavoidable perplexity* is justified and it encourages us to "do something".

Advantages and disadvantages of possibility 1 (the present) include: the doctor feels justified in not investigating any further; there are fewer interventions for patients, but perhaps some cases are erroneously labelled as NCGS. To what extent does recognising NCGS as a valid diagnosis expose us to an additional risk?

A case could arise in which correct diagnosis of a patient with a severe disease is delayed because that more severe condition may also respond (though only transitorily) to the gluten-free diet, as in, for example: patients with an inflammatory intestinal disease, intestinal lymphoma, pancreatopathies, other food intolerances or intolerance to biliary salts, among other causes. In all these cases, a certain time after presenting a false response to the gluten-free diet, the patient would once again present the initial symptoms (diarrhoea, abdominal pain, etc.).

At this juncture, the point of view of the general practitioner will probably vary from that of the specialist. To the former, the delay in these (few) cases seems hardly to be significant in terms of prognosis. In contrast, to the gastroenterologist, the higher prevalence of the more serious disorders could justify requesting further, more sophisticated diagnostic tests; and this position is supported by an exhaustive review of 189 cases of NCGS of which almost 30% were reclassified as CD (Molina-Infante et al. 2015).

This difference in behaviour between the general practitioner and the gastroenterologist is fully supported by Bayes' theorem: a low "pre-test" prevalence inevitably leads to a greater rate of false positives, which means that the general practitioner should be less interventionist in order to reduce the iatrogenesis associated with additional interventions, (what has come to be called *quaternary prevention*: protection from overmedicalisation). We can conclude (in keeping with Bayesian epistemology) that the degree of diagnostic effort and precision depends not only on the patient's clinical presentation, *but also on the clinical context.*

However, from the invariantist position, it could be argued that there is sufficient semiotic and biological data on each patient for us to label them with great precision in the current diagnostic taxonomy. Using the clinical context as an excuse to apply a strategy of *avoidable perplexity* would only be justified by economic considerations, or a lack of human resources, but never because we have encountered an insurmountable obstacle and even less so a "serious" epistemological option. So, good clinical practice would not allow these "loose ends" to be left unattended and would aim to provide a precise diagnosis for every patient, without giving way to a low pre-test prevalence.

If we look back for a moment to the start of this article, we said that there are two main (nosological) tendencies when it comes to defining a disease or illness: essentialism and pragmatism. Now we perceive another interesting division *in relation to how we apply this knowledge*: contextualism *appears in opposition to* invariantism.

If we combine these options, we can define four typologies. The most drastically opposed two (Table 2) would be:

(a) Invariantist essentialist professionals: patients can be diagnosed with diseases that have well-characterised biological markers, and that have a specific

Table 2 Nosological options for the understanding of diseases

	Essentialist: diseases have a "core" essence that defines and distinguishes them	Pragmatist: many illnesses and disorders can only be defined by means of functional and flexible criteria
Invariantist: We can (almost) always arrive at precise diagnoses	Diseases have well-defined illness scripts and are anchored in biological evidence: Patients can and must be precisely diagnosed	Although we group patients together at the functional level, if overall they exhibit similar behaviour, then we are faced with an invariant phenomenon that we should define (and act accordingly)
Contextualist: a precise diagnosis is not always possible or desirable, it depends on the context	Important diseases can adopt idiosyncratic forms but we can predict the risk of suffering from them by adopting a Bayesian approach	Illnesses are defined through fundamentally probabilistic criteria, which sometimes leads to scenarios of avoidable perplexity

treatment and prognosis. There are disorders or conditions that have not achieved this degree of definition, but—from this perspective—it is a question of time before our knowledge of them does reach such clarity. This position would be that of specialist physicians who move in the ambit of highly specialised hospitals. The use that they make of state-of-the-art technology confirms to them their epistemological position. From this perspective, they would recommend "*the minimum possible number of diagnoses of NCGS and maximising the possibility of biopsies after exposure to gluten*".

(b) Contextualist pragmatist professionals: there are no diseases, just ill people. Each illness acquires an idiosyncratic form in each patient which makes it necessary to adapt the diagnostic and therapeutic strategies to each case. There are advances in biology that undoubtedly are a great help, but it is always necessary to reinterpret the illness script in accordance with the social and personal context. Patients can be grouped together via functional criteria, for descriptive purposes, and that helps us to choose and mix personalised therapeutic options freely. This position would typically be held by general practitioners, whose task with greatest added value is to distinguish between many psychosomatic complaints and those few that may represent a real risk to the patient's life. The use of nomograms and risk charts provides them with the necessary Bayesian approach. From this point of view, *it is unadvisable to go ahead with a diagnosis of CD when the antibodies and HLA haplotype are negative.*

Between these two opposing epistemological poles we find the other two less extreme typological variants:

(c) Contextualist essentialist professionals: there are "perfect" models of diseases, but the clinical and psychosocial contexts introduce variations that mean we have to personalise risks and act in an individualised manner (risk charts, nomograms, etc.). Consequently, sometimes we will have to give up the search for diagnostic precision because "we have done all we can" with the means we have at our disposal. This position is typically defended by specialist doctors who work in external consultancies and small regional hospitals with considerable pressure on resources. From this perspective, it is desirable to refine the semiotic perception of each case as far as possible in order to select those patients for who the benefit-to-risk relationship makes it worthwhile pursuing the most costly complementary explorations. In an imaginary dialogue with a specialist from a large tertiary hospital our specialist from a small regional hospital would recognise that on occasions the diagnosis is not taken as far as it could be, but may add: "*if we didn't work that way the National Health System would become unviable*". And to round things off: "*I agree that I have some patients whose diagnostic process I could pursue further, but I act according to a curve of diminishing returns: when I perceive that the cost or effort involved will only result in miniscule therapeutic utility, I bring my diagnostic efforts to an end and move on*".

(d) Invariantist pragmatist professionals: many people may not receive a precise diagnosis, but even so, when large groups of patients are analysed, we get things wrong on very few occasions. We do not know what fibromyalgia consists of, but the prevalence of widespread pain throughout the body is constant and the same across very different societies. It will take us more or less time to classify this group of patients, but they are there and will continue to be. The further we move our perspective outwards from individual cases and towards getting a big, global picture, the clearer things become. Things "are what they are", although we constantly have to accept definitions of cases that are elusive, especially for psychopathological cases. This epistemological position makes the healthcare administrators who have to back some or other public policies feel safe. In the case at hand, the presence of a series of patients with NCGS would be the necessary and sufficient guarantee to award them entity status, so long as there is temporal and transcultural consistency.

Healthcare professionals may adopt these different perspectives at different stages of their life. Choice of the group we belong to can be influenced by our workload, the means we have at our disposal and whether we see it to be necessary to define healthcare policies. We leave it to the reader to apply this typology to other cases, for example the prevention of prostate cancer through early diagnosis, or widespread pain throughout the body.

This analysis would be incomplete if did not look further afield in order to analyse how the scientific community and broader society in general react to this debate, and the fashions and trends it gives rise to.

Have We Left Ourselves without a Model of Disease?

The classic model of disease proposes the existence of a disorder which—with few variations from one generation to the next—produces a uniform change in the subject who suffers from it, whose clinical manifestations are specific and which can be defined by a set of criteria (Adriaens and de Block 2013). The case we are concerned with here, NCGS, and food intolerances in general, do not fit in well with this concept of disease.

In such cases we cannot speak of a specific genetic defect, or of a uniform set of symptoms. This is why it is useful here to introduce the concept of a multifactorial disease (Table 3).

Keller and Miller (2006) use the following metaphor to explain the genetic component of this type of disease. First, imagine a river basin that collects water via many small streams and this all comes together and ends up in one powerful river. Although there are different genetic defects—these would be the streams and rills that flow down from the mountains—in general, they do not lead to important changes in the flow of the large river. And when an important change does occur, it is usually due to some external factor, a lack of rain, for example, which affects some streams and not always the same ones. This model, if it is correct, predicts that depression, NCGS and many food intolerances do not stem only from a single molecular defect, or even from the presence of a set of genes that we can identify in a stable way. We would rather be dealing with groups of genes with specific functions that change in specific environmental situations (epigenetic changes).

Now consider that human perception is prepared to interpret those phenomena that best fit models of perfection, which we have said pertain to deterministic systems (or what we humans understand as such). If this law was true, we would be prepared to isolate diseases that match a model of perfection, to the detriment of detecting others (for example, Mendelian diseases to the detriment of multifactorial ones). This would certainly seem to be what the history of our medical taxonomy confirms. We could call this phenomenon *perfection masking*.

Applied to food intolerances, our cognitive apparatus initially selected and labelled those patients which matched the classic model of disease. NCGS was left

Table 3 Mendelian, polygenic and multifactorial diseases

Mendelian disease	A well-defined genetic defect that, since it affects a chromosome or an allele, produces a specific disease or disorder
Polygenic disease	A well-defined combination of genes that almost always produces a disease, but to do so requires there to be a certain interaction with the environment. This would be the case of the majority of cancers, some neurodegenerative diseases, and some psychiatric diseases such as schizophrenia, CD, some dyslipidaemias, etc.
Multifactorial disease	A variable combination of genes which only sometimes leads to disease, and almost always through interaction with predispositional factors in the environment (epigenetic changes). This would be the case of many mental diseases, learning disorders, NCGS, etc.

out of that initial understanding and could not form part of the model of perfection, in this case of CD, as it was *masked and waiting for pragmatist professionals to defend it and demonstrate its existence through more lax, less "perfect" criteria*.

In the case of NCGS, healthcare professionals realise perfectly well that they are on a slippery slope; that this diagnosis is less well grounded than if they had managed to demonstrate the presence of CD. This is the same situation that they experience when they give a diagnosis of polymyalgia rheumatica (PMR), or fibromyalgia; the tension that these *weak* diagnoses causes is lessened if the response to the specific treatment is *strong*. In fact, NCGS (just like PMR) has well-defined treatment that improves the situation in many cases, thereby representing a great advantage over fibromyalgia. A more in-depth metaphysical discussion of this issue can be found in Borrell and Estany (2016).

Is this where the story of NCGS ends? Can we adapt to this new type of diagnosis? It would be a mistake to argue that healthcare professionals must abandon the model of perfection and open themselves up to a pragmatist model. The tension between models with an essentialist basis and those with a pragmatist base can lead to positive outcomes. On the one hand, the pragmatist position allows elusive phenomena to be identified, such as those that are initially hidden from view by what we have called masking. On the other hand, the essentialist position makes it difficult for healthcare professionals to imagine diseases that cannot be verified, as it requires a model of illness anchored in biological or physiopathological evidence.

The recurring discussions and arguments in psychiatry when it comes to defining entities (such as "somatoform disorders", "eating disorders", or "personality disorders", among many others) illustrate this tension (García-Maldonado et al. 2011; Zachar 2014). But in other specialties we can also find examples, as in the case of *Helicobacter pylori* and peptic ulcers, and the transition that the connection underwent from a model of "circular causality" to a classic model of *conditional causality*.

The appearance of NCGS obliges us to reconceptualise food intolerances as a spectrum of disorders rather than as specific discrete diseases. It invites us to find new molecules that can explain and anchor IIS and it also invites us to reconsider the relation between food, digestion and mental activity (certain cognitive and behavioural disorders). The conceptual map becomes notably richer and we generate promising new hypothesis: Do similar effects occur with other cereals in humans or other species? Are similar molecular mechanisms involved? Can we isolate antibodies with similar features whose mere presence should make us suspect that there are other food intolerances at work? Via what pathways do certain nutrients influence our cognition? And so on.

The pragmatist–essentialist tension therefore opens up a whole field of extremely stimulating research.

Food Fashions: From Science to Magical Wonder

When we consider how society reacts to this debate, we could talk of several levels of interpretation.

- "*NCGS is one more entity—among other possibilities—that could explain my digestive problems*". In this case, the patient would be making a non-judgemental and prudent interpretation of a news story from the media.
- "*NCGS shows that there is a series of toxic substances (gluten, lactose, fructose, etc.) that if eliminated from the diet have beneficial effects for the whole population*". In this case, a careless overgeneralisation is being made from a specific scientific finding. People who followed this line of thought as a recommendation would miss out on the advantages of certain foods and they would complicate their diet for no reason while also making it unnecessarily more expensive, and on occasions this could even become dangerous.

It is certainly the case that there is a tendency for commercial interests to invent new diseases, taking advantage of physiological variations, processes of aging that are to be expected or certain discoveries. Undoubtedly there is an enormous volume of potential business where diets are concerned: 7% of German consumers claim to be on a gluten-free diet, and this figure increases to 11% among young people aged 16–24 (the millennium generation) (Free From Food Expo 2016).

At the "Free From Functional Food Expo 2016" (Amsterdam) there were 250 stands and the event received more than 4000 visitors. The Mad Gluten Free event in Barcelona in 2016 was visited by 9000 people. "Free from" foods can now be found on the shelves of almost every supermarket and are no longer confined to specialist shops.

In the light of all this, the question we should be asking is whether these are positions that are supported by our current best scientific knowledge. In fact, the Mediterranean diet is the only one backed by scientific evidence of its health-related benefits, together with the evidence of the negative effects of consuming too much salt, red or processed meat, and foodstuffs with added sugars (Estruch et al. 2013). Beyond these few recommendations, however, we enter a magical realm of wonder captured so well by the famous slogan "*we are what we eat*", or the more practical: "avoid *cancer by following this diet*".

All of this brings us back to the model of a "natural life". From its beginnings, Western culture attached great importance to the term "nature" (physis); and just as the ancient Greeks did, we continue to believe that what is natural is good, healthy, beautiful and so on. In the absence of religious myths that remove the fear of death, many people turn their sights to "the natural", as if a "bio" product could promise them a peaceful life free of affliction (Borrell 2017).

As if that was not enough, in the case of NCGS, the range of symptoms that the syndrome can produce runs the risk of encouraging us to recommend a gluten-free diet to a large proportion of the population. If that were to happen, over-diagnosis and over-treating would be unavoidable. Perhaps in a rather exaggerated way, the first studies that warn of the negative consequences of a gluten-free diet are starting to crop up (Lebwohl et al. 2017; Vici et al. 2016). On the positive side, there are thousands of patients who may be sensitive to gluten but do not know it and for them, NCGS could be very good news.

Despite all this, this debate certainly is not new. Each novel piece of knowledge that bursts onto the medical scene obliges us to take a long sober look at its practical and prudent use. Misuse (or simply *overuse*) of *good news*, in medicine, *almost always causes victims.*

Acknowledgements This work was supported by the Spanish Government's Directorate-General for Science Research and Development via research project: FFI2017-85711-P Epistemic Innovation: The Biomedical Sciences Case.

This work is part of the consolidated research network "Grup d'Estudis Humanístics de Ciència I Tecnologia" (GEHUCT) ("Humanistic Studies of Science and Technology Research Group"), recognised and funded by the *Generalitat de Catalunya*, reference 2017 SGR 568.

References

Adriaens, P. R., & de Block, A. (2013). Why we essentialize mental disorders. *Journal of Medicine and Philosophy, 38*, 107–127. https://doi.org/10.1093/jmp/jht008.

Biesiekierski, J. R., Newnham, E. D., Irving, P. M., Barrett, J. S., Haines, M., Doecke, J. D., Shepherd, S. J., Muir, J. G., & Gibson, P. R. (2011). Gluten causes gastrointestinal symptoms in subjects without celiac disease: A double-blind randomized placebo-controlled trial. *The American Journal of Gastroenterology, 106*, 508–514.

Borrell, F. (2017). La hipótesis de una vida natural. *Folia Humanística* (5), 32–41. Accessed May 2017, from http://revista.fundacionletamendi.com/index.php/ficha/101/La-hiptesis-de-una-vida-natural

Borrell, F., & Estany, A. (2016). Construcción y refutación de nuevas enfermedades. El debate entre esencialistas y pragmáticos desde la perspectiva de la epidemiologia social. En: A. Estany & A. Puyol (Eds.), *Filosofía de la epidemiologia social* (pp. 119–142). Madrid: CSIC. Plaza y Janés.

Brown, P., & Levinson, S. (1987). *Politeness: Some universals in language usage*. London: Cambridge University Press.

Catassi, C., & Fasano, A. (2010). Celiac disease diagnosis: Simple rules are better than complicated algorithms. *The American Journal of Medicine, 123*(8), 691–693. https://doi.org/10.1016/j.amjmed.2010.02.019.

Catassi, C., Bai, J. C., Bonaz, B., Bouma, G., Calabrò, A., Carroccio, A., Castillejo, G., Ciacci, C., Cristofori, F., Dolinsek, J., et al. (2013). Non-celiac gluten sensitivity: The new frontier of gluten related disorders. *Nutrients, 5*, 3839–3853.

Catassi, C., Elli, L., Bonaz, B., Bouma, G., Carroccio, A., Castillejo, G., & Dieterich, W. (2015). Diagnosis of non-celiac gluten sensitivity (NCGS): The Salerno experts' criteria. *Nutrients, 7* (6), 4966–4977.

Díaz Marugán, V., Magallares García, L., Fernández Caamaño, B., Alcolea Sánchez, A., Alonso Canal, L., & Polanco Allue, I. (2013). ¿Puede ser el gluten perjudicial en pacientes no celiacos? *Evidence Pediatrics, 9*, 1. Accessed May 2017, from http://www.evidenciasenpediatria.es/DetalleArticulo/_LLP3k9qgzIh7aNQBiadwmcEHytVoiLMhc1TXy9MJ0TRUR3wYDDHdaLPTzCm2cl3v2NcjxqC8gLocSzbokkM2Lg#articulo-completo

Estruch, R., Ros, E., Salas-Salvadó, J., Covas, M. I., Corella, D., Arós, F., et al. (2013). Primary prevention of cardiovascular disease with a Mediterranean diet. *The New England Journal of Medicine, 368*, 1279–1290 Accessed May 2017, from http://www.nejm.org/doi/full/10.1056/NEJMoa1200303#t=article

Free From Food Expo. (2016). *Interviews*. Official catalogue Free from/Functional Food 2016 RAI Amsterdam. Accessed May 2017, from http://www.freefromfoodexpo.com/pdf/2016-catalogue-interviews.pdf

García-Maldonado, G., & Atenógenes, H., González, S., Llanes, A., Sánchez, I. G., & El DSM, V. (2011). Luces y sombras de un manual no publicado. *Retos y expectativas para el futuro Salud mental, 34*, 367–378. Accessed May 2017, http://www.scielo.org.mx/pdf/sm/v34n4/v34n4a9.pdf

James, W. (1997). *Pragmatismo*. Madrid: Ed.Santillana.

Keller, M., & Miller, G. (2006). Resolving the paradox of common, harmful, heritable mental disorders: Which evolutionary genetic models work best? *Behavioral and Brain Sciences, 29*, 385–452.

Lebwohl, B., Cao, Y., & Hu, F. B. (2017). Long term gluten consumption in adults without celiac disease and risk of coronary heart disease: Prospective cohort study. *British Medical Journal, 357*, j1892. https://doi.org/10.1136/bmj.j1892 Accessed May 2017, from http://www.bmj.com/content/357/bmj.j1892

Molina-Infante, J., Santolaria, S., Sanders, D. S., & Fernández-Bañares, F. (2015). Systematic review: Noncoeliac gluten sensitivity. *JO Alimentary Pharmacology & Therapeutics Aliment Pharmacol Ther, 41*(9), 1365–2036.

Peirce, Ch. (1877). The fixation of belief. *Popular Science Monthly, 12*, 1–15. Accessed May 2017, from http://www.peirce.org/writings/p107.html

Schmidt, H. G., & Rikers, R. M. (2007). How expertise develops in medicine: Knowledge encapsulation and illness script formation. *Medical Education, 41*(12), 1133–1139.

Schön, D. A. (1983). *The reflective practitioner: How professionals think in action*. New York: Basic Books.

Van Schaik, P., Flynn, D., van Wersch, A., Douglass, A., & Cann, P. (2005). Influence of illness script components and medical practice on medical decision making. *Journal of Experimental Psychology Applied, 11*(3), 187–199. https://doi.org/10.1037/1076-898X.11.3.187.

Vici, G., Belli, L., Biondi, M., Polzonetti, V., et al. (2016). Gluten free diet and nutrient deficiencies: A review. *Clinical Nutrition. 5*(6), 1236–1241. https://doi.org/10.1016/j.clnu.2016.05.002. Accessed May 2017, from https://www.ncbi.nlm.nih.gov/pubmed/27211234

Volta, U., Caio, G., Tovoli, F., & De Giorgio, R. (2013). Non-celiac gluten sensitivity: Questions still to be answered despite increasing awareness. *Cellular & Molecular Immunology, 10*, 383–392. https://doi.org/10.1038/cmi.2013.28; published online 12 August 2013. Accessed May 2017, from https://www.ncbi.nlm.nih.gov/pmc/articles/PMC4596973/

Zachar, P. (2014). *A metaphysics of psychopathology*. Cambridge: MIT.

Progress in Biomedical Sciences and Raw Data: Ethical Dilemmas

David Casacuberta and Simone Tassani

Abstract How do we assure that a scientific result is sound, that no fraud has been committed and we do guarantee scientific progress in this new era of big scientific data? In order to guarantee reproducibility, especially in the biomedical sciences one need to give access to third parties to analytical datasets (either raw or processed), software used, code developed by the researchers, a very detailed description of the methodology as well as all the relevant metadata. We want to argue that that sharing data is mandatory in order for science to progress. Each published paper should not be seen as an untouchable revealed truth, but as an opportunity of dialogue, confrontation and verification. This can be done only if all raw data are shared and each researcher has the right tools to reproduce an experiment.

Keywords Scientific progress · Reproducibility · Epistemology of big data · Raw data

The Pursuit of Reproducibility and Scientific Progress

How do we assure that a scientific result is sound, that we can consider it at least provisionally true (Popper 1963)? This has been and still is one of the major problems in philosophy and methodology of science, and just presenting the different solutions proposed by philosophers would take several thousand pages. Nevertheless, most philosophical literature, as well as basic intuition of scientists, (their "spontaneous philosophy" as twentieth century French Althusser call it, see Macherey 2009) agree in the key importance of reproducibility.

D. Casacuberta (✉)
Philosophy Department, Universitat Autonoma de Barcelona, Barcelona, Spain
e-mail: david.casacuberta@uab.cat

S. Tassani
Department of Information and Communication Technologies, Universitat Pompeu Fabra, Barcelona, Spain

J. Vallverdú et al. (eds.), *Philosophical and Methodological Debates in Public Health*, https://doi.org/10.1007/978-3-030-28626-2_6

According to the U.S National Science Foundation "reproducibility refers to the ability of a researcher to duplicate the results of a prior study using the same materials as were used by the original investigator. That is, a second researcher might use the same raw data to build the same analysis files and implement the same statistical analysis in an attempt to yield the same results [...] Reproducibility is a minimum necessary condition for a finding to be believable and informative." (Bollen et al. 2015). From a philosophical perspective there is a lot more to say about this, specially when considering the relationship between reproducibility and truth (See for example, Goodman et al. 2016 which argue that current insistence on replicability can be viewed as setting aside a proper scientific focus on truth). Collins and Tabak (2014) shows how reproducibility is not a clear cut process, as it usually implies to describe how the experiment was designed, reported, analyzed, interpreted and corroborated.

Also, demanding reproducibility, is more a methodological and ethical postulate than an actual condition for acceptance. Science publications keep soaring. As stated in Sintef (2013) 90% of all the data related to scientific results were obtained after 2010. Also, it is difficult to build a scientific career reproducing other people's results, as novel, original ones have a deeper impact in a researcher's CV. Nevertheless, that doesn't imply that reproducibility is not key for science development, otherwise, we'd be falling for the naturalistic fallacy (Moore and Baldwin 1993).

Fortunately, we can skip that higher level of argumentation and just accept—for the sake of the discussion—that reproducibility is a very desirable attribute to consider in a biomedical debate. A reproducible research is not automatically a good research, but non-reproducible research is more likely to be considered opinion and not science.

In order to guarantee reproducibility, especially in the biomedical sciences one need to give access to third parties to analytical datasets (either raw or processed), software used, code developed by the researchers, a very detailed description of the methodology as well as all the relevant metadata. That means, as we will discuss later, that reproducibility is a difficult process. In medical and bio-medical research, biological variability impedes exact replication of data, no matter how clear the methodology is described. Studies are today commonly based on patient specific approaches or on complex statistical data analysis, and cannot be replicated without a direct access to the raw data of the study. This, in medical research, would mean to share all the data related to clinical trials, surveys, medical images and even software developed.

When replicability is sought, as can be seen in several comprehensive studies, reproducibility level is quite low. According to Begley and Ellis (2012), when Bayer tried to replicate most quoted results on cardiovascular diseases, oncology, and women's health—published in peer reviewed, high impact journals, they could only reproduce a quarter of them. Prinz et al. (2011) describes how when the American drug company Amgen tried to reproduce results from 53 high impact cancer research papers they could only replicate 11% of the results.

Most often, when trying to reproduce results, researchers usually found that there is not enough information in the paper to really repeat the experiment or the analysis. For example, Vasilevsky et al. (2013) describes how half of 238 papers from 84 different journals didn't provide all the resources required to really verify the results.

Then add fraud, sloppy statistical procedures, poor training of the researchers in experimental methods, lack in paper of basic procedures like randomization or double blind, doctoring the data due to the publish or perish pressure, lack of space in the journal to describe complex procedures, Megabytes or even Gigabytes of data difficult to share, researchers hiding key information in order to keep a competitive value, and it is easy to understand how difficult is to achieve replication of scientific results nowadays.

Scientific progress clearly resents from it. Retractions of supposedly relevant results are on the rise, getting a tenfold increase in the past decade (Cokol et al. 2008). Lack of relevant data also makes peer reviewing quite inexact and prone to errors. Consider the prank developed by John Bohannon, a biologist from Harvard, which, Sokal style (Sokal and Bricmont 1997), was able to publish a fake paper full of errors that described how a drug obtained from lichen was able to treat cancer cells. (The Scientist 2013). Also, when 420 reviewers of BMJ were invited to evaluate a paper that included eight deliberate errors, of the 221 that answered, almost 20% couldn't find any errors at all, and the median number of errors spotted was 2 (Wakefield et al. 2010).

Scientific Fraud and Doctoring Results

Technology makes everything easier, even committing frauds, and they are more difficult to detect. Reproducibility is most of the times the only weapon to establish such frauds. Jan Hendrik Schön is a German physicist that in 2000 and 2001 was able to publish 27 articles (in only 2 years) in very high impact factor journals like *Science* and *Nature*. Reviewers were powerless, and it was only thanks to the attention of the scientific community trying to replicate his results that the fraud was finally discovered. It was a huge scandal, but the actual impact over the community was minor since the period from publication to discovery of the fraud was short.

Yoshitaka Fujii is a Japanese researcher in anesthesiology who in 2012 was found to have fabricated data in at least 172 scientific papers. Fujii committed these frauds during his whole career. Of these 172, 126 papers were determined to have been "totally fabricated". Those 19 years of frauds had many dark consequences, like unhelpful clinical trials repetitively tested over patients that underwent surgery causing them problems and suffering. It was difficult, time consuming and dangerous to falsify Fujii's studies. Once again it was only due to the attention of the community that was trying to replicate his results if finally his fraud was discovered.

Fortunately, fraud very seldom happens in science. For example, according to (US HHS 2011) in that year, only 12 cases of fraud were investigated. But that doesn't mean that fraud is irrelevant, as the former examples of Schön and Fujii clearly show. Also a meta-analysis of fraud surveys showed that despite only 2% of the researchers accepted to have committed some sort of fraud, a worrying 14% reported to have seen such type of unethical behavior in some colleagues (Hughes 2014).

Sharing Raw Data to Help Scientific Process

Fortunately, the scientific establishment is well aware of such difficulties and there are several initiatives to address them. Lack of access to results is dealt with the concept of open access, the practice of providing unrestricted access via the Internet to peer-reviewed scholarly journal articles, or by giving space to publish negative results. Irregularities with the review system is tackled with non anonymous peer review (Casacuberta and Estany 2003). And there is a special focus on reproducibility, like the newly formed Reproducibility Initiative. Here, scientists can hire an independent laboratory to get their work verified (Hughes 2014).

Equally relevant, the NIH is giving a better formation for researchers on experimental methods and statistical analysis, as well as a development of a checklist to facilitate reproducibility (Collins and Tabak 2014). Also by the NIH there is the Data Discovery Index (DDI) designed to facilitate researchers to share and download unpublished, primary data. Some journals like *Perspectives on Psychological Sciences* are starting to include specific sections on replication and verification. There is even, as described in Hughes (2014)—a law in the US that makes mandatory to publish your data if your research has been financed with public money.

Much more common is cherry-picking, reporting data that supports a desired outcome, or doctoring the data, eliminating some subject or trial that make the results look a lot less relevant. The number of experimental observation is often low and one single specimen can make the difference from a successful and a meaningless statistical analysis. Biological data are very easily alterable. It is extremely easy to modify some data, creating a probable result, without any possibility to verify them. This is due to biological variability. A study performing exactly the same procedure of an already published work will never produce the same results. In the best case results will be similar. A biological or medical study always involves a huge number of variables, related to factors difficult to control, like the lifestyle of patients. Scientists usually have direct control only over few of them and are more or less able to monitor the remaining ones. This is the main difficulty and the beauty of studying biological systems. At the end, changing one variable can completely change the results of the study. In this scenario it is quite easy for someone with bad intentions to doctor, or even fabricate new sets of data.

Scientific Career Versus Raw Data

According to Article 27 of the Universal Declaration of Human Rights, "everyone has the right to the protection of the moral and material interests resulting from any scientific, literary or artistic production of which he is the author" (Assembly 1948). Data are property of the researchers until they are published. This is an extremely delicate topic and it is probably the main "drawback" of an OA repository. To develop the right protocol can require months of study, and then to implement it, find a willing significant number of patients and follow them to check the effect of surgery or drugs is a job that can take up to 5 years or more.

It shouldn't surprise us then that scientists in general do not like the idea of sharing data that difficult to obtain to the community. Those data will probably lead to several publications as well as future funding. If data is shared it would give other teams all over the world the same advantage of the original team, with the difference that they will not need to spend years in collecting and analyzing the data. So, by publishing raw data, researchers will lose the IP over them and everybody else could use these data and publish scientific papers based on them, even before the original owner of the data.

This is a critical point since it can potentially lead other competitors to publish more and faster, reducing the advantage of the researcher and therefore their reputation. Even if a scientist doesn't care about prestige, she perfectly knows that the evaluation of her productivity as a researcher is mainly based on publication of innovative studies in top scientific journals. Losing her IP over those complex data will probably reduce her productivity of innovative studies (because someone else could do it before her) and therefore reduce her impact among the community, her credibility and, finally, her funds.

What Should Weight More?

This presents an ethical dilemma. What is more relevant to scientific progress? To share raw data so scientific research can be verified and replicated? Or to let things as they are now, so a scientific team can hold intellectual property over their raw data?

Let's recapitulate what we have discussed so far:

If we consider open access to raw data more valuable, then we will have tools to avoid and prevent scientific fraud. As we have seen before, raw data access is a key element to discover and expose fraud, and as we also described above, despite fraud is not a huge issue, it is neither negligible.

We already mentioned how the main problem though, is not fraud but sloppy research. Again, if raw data is published, those mistakes could be more easily spottable. Scientists, knowing they can be more easily audited, will be more careful in the way they analyze their data and present their results and the community will gain a strong tool to take care about itself.

As we have argued in the former section, when raw data is released other scientific teams can re-use it. This nowadays is bad for the scientific team that generated the data, but it is clearly good for scientific progress per se, as more teams will be able to work on the same problem using that relevant data.

Finally, let us not forget that this ethical problem goes beyond a mere internal issue in how scientific method is applied. Citizens are starting to wonder whether science is or isn't a good investment for their taxes. In October 2013 an article published on The Economist with the title "Trouble at the lab—unreliable research" (The Economist 2013) underlined the problem of lack of repeatability of experiments that give space to extravagant interpretations of statistics, misbehaviors and frauds. More recently, in July 2014 The Wall Street Journal wrote "The Corruption of Peer Review Is Harming Scientific Credibility" (Campbell 2014) in which limits and frauds related to peer review are underlined. Today the credibility of the scientific community is a boost for many industrial applications that can claim the support of a scientific approach to their products. Tomorrow this might not be the case. Reputation of the scientific community is at risk and we all need a big effort to change the direction since, even if is relatively easy to lose credibility it is extremely difficult to gain it back again (Bourne and Barbour 2011).

Releasing the data as soon as possible have clear implications for the general public. The more data researchers have, the more research advances and the easier to produce new drugs and protocols to tackle illnesses. There are also economic advantages in re-using raw data for the private sector, either in the biomedical industry or consulting services.

Now let's us consider the drawbacks that a scientific team may have to assume if they release their data before a paper is published.

In the more extreme situation, scientists forced to publish the raw data publicly before a paper is published, it would be extremely bad for the original scientific team, as a better, or just more numerous team could take the data and find the same results than the original team and publish it before.

Another less extreme situation makes it mandatory to add the raw data when the paper is submitted to a journal, so reviewers can check it before deciding to publish or not. This is not so dramatic for the original scientific team, but still, an unethical reviewer could decide to use the raw data for his own research while stopping the publication of the original team and then publish the results first in another journal (Casacuberta and Estany 2003). It would also make the reviewing process a lot slower, therefore damaging scientific progress.

Our proposal therefore, will be the following: In our society, science has an enormous importance. Science has also turned a lot more complex as well as competitive. This makes highly important to distribute raw data in an open access manner, to assure reproducibility and therefore, scientific progress, as well as more engaged and respectful view of science from the general public. In order to assure that basic intellectual property rights of scientists are assured, those raw data should be made public only when a relevant paper based on those data is published and authorship of the data is clearly linked to the original authors.

Some Objections to Our Thesis

One could point out to the fact that sharing raw data is easier said than done. Raw data are rarely published, especially when they are too complex to fit in a table or a 2D graph. A multidisciplinary approach to biomedicine requires several specialists to work together. Their data are extremely different and their degree of complexity can vary. The development of a global repository for data sharing is twice challenging from the technical point of view. It requires developing new infrastructures able to handle very complex results, like complex three dimensional simulations, but must also create a relation between different results from different field of study.

Nowadays, it is impossible to publish on paper all the data related to a specific experiment. It is not only a problem of space, it is literally impossible to present complex multi-dimensional analysis in the two dimensional space of a classic .pdf file.

It is not enough to upload data in order to share them with the community. Taxonomy and ontology studies must be performed in relation to each kind of data in order to create a structure that can receive the data and organize them in an accessible way. All the structure should be developed in cooperation with different users and manufacturers of device used in order to develop the data. It is not surprising if many problems still need to be solved in order to develop such a kind of repository.

However, returning to the concept of the naturalistic fallacy, we are not discussing about how feasible an ethical practice is, or how extended is right now, but how scientific research should be like. However, this doesn't imply any ethical paradox as we are not asking people to assume something impossible in practice, like a pure Kantian ethics or the Boddhisattva vow, but to tackle a technological problem. Also, if sharing raw data becomes a common scientific practice, it would be easier to find technological expertise as well as budgets to implement such protocols. After all, if digital entertainment companies can offer their subscribers full movies and TV series on demand, in high definition, there is no technical reason for not being able to do the same with scientific raw data.

Another relevant objection is that instead of helping scientific progress such measure would probably reduce it. If scientists are forced to release the data they have been accumulating for months, even years, then they lose any strategic advantage. So, excellent teams will have a lot less motivation to look for novel, relevant data and free rider teams would prosper, using other people's results for their own good without giving anything back.

This is the main reason we propose that data should be released only after the publication, and that the original team should keep the IP over it. In order to avoid the drawback described here, metrics of scientific results should be redesigned to assure that if a team B publishes some results based on data originally obtained by team A, then team A gets recognition and reputation out of it, even if they haven't participate in the writing of the paper.

A more philosophically oriented mind may point out that, ethically, not doing bad things is different from forcing people to do good things. In spirit of Judith Jarvis

Thompson's argument for a right to abortion (Thomson 1976) we should observe that our moral norms are mostly oriented towards the former: we consider that hurting other people is a bad action and even have some laws to punish those that practice it, but no laws force us to spend our lives in hospitals helping people that are sick. One can volunteer to do that, and there are good infrastructures to do so, and we all congratulate when that happen, and may even give money to NGOs doing that, but nobody is forced to do that. We can rapidly relocate that argument to our context and say that what an ethical system should do when considering scientific progress is to stop fraud, but not forcing scientists to be altruistic and release their raw data.

We agree with the general idea behind it. As it happens with open access papers, which is just an option—on the rise—for scientists, we expect that an open access to raw data first become an option which will grow as more scientists realize that it is the best option for all.

Nevertheless, our main worry is not stopping fraud per se, but guaranteeing scientific progress as well as improve the image of scientific research among the general progress, so, in search for a greater good, however not becoming mandatory, we think it would be good if scientific community starts to consider those papers published under our model to be more reputable and scientifically sounder, and therefore make this option the most relevant for a scientific team. If this proposal only becomes a curious second option for second level results all the problems that we have been discussing so far will not be solved. As we have described in section "Sharing Raw Data to Help Scientific Process" this is already taking place: the DDI database, for example, is designed so as to give credit to the author of the data.

Then we have the objection of publicly releasing data about people, putting their privacy in jeopardy. This is a very relevant issue, without a doubt, but it is not really confined to our proposal. Massive databases on patients are routinely accessed and distributed for scientific research. Adding those to a public repositories only would make them more accessible. Researchers are already using strong cryptography (Aggarwal and Philip 2008) to assure that anonymized data cannot be re-identified again, and despite former attacks that revealed identities in the past (Sweeney 1997) there seem to be already good protocols to assure that medical records are completely anonymized (Barth Jones 2012).

Conclusions

The need to share raw-data, at least to improve the quality of the research, is becoming obvious. We propose to create reputation mechanisms to force the publication of raw data after the study is accepted for publication by a scientific journal. In order to publish a study on a scientific journal, the researcher must stand a process of peer review during which the paper is revised by two or more reviewers. Once the reviewers decide for publication of the study, the editor of the journal could ask the researcher to share their raw data on a specialized repository. This easy solution would solve the majority of the presented problems, as explained below.

Our proposed solution would keep the IP of the researcher over their data until the publication of their results. Since the reputation of researcher is mainly based on publication of innovative studies in scientific journals, in order to protect the IP of the original owner we should recognize the paternity of the data and to cite the original owner.

An agreement on the kind of conduct to keep for the publication of the raw data does not automatically solve the technological problems related to the practical realization of the OA repository. Nonetheless, once the direction is set, the whole community will have the chance to focus only on the practical problem related to the development of the repository. The necessary technology already exists. A synergic effort of the scientific community is all we need to develop this kind of repositories.

Such an agreement would help the development of scientific progress: fraud and doctoring data would be a lot less common, because they will be a lot more easily detected. It is impossible to reduce to zero the possibilities of data fabrication. Nonetheless, the introduction of this procedure, as a routine for the publication, will increase the technical difficulties in data fabrication. Fabricate data becomes more difficult when the results of one researcher must be coherent with the results others. Sharing raw data with the whole community makes this process even more difficult. It will be easier to verify the results presented by authors in their papers and replicate their studies.

When individual fraudsters are considered, they will be a lot more easily detected, as it will be extremely difficult to share falsified raw data keeping the coherence to the related data shared by the colleagues of the same laboratory.

When the public opinion is considered, the proposed solution will increase the sharing behavior and the idea that the scientific community cares about the integrity of the researchers' conduct and it will control (and potentially punish) eventual misconducts.

The community is already moving toward the direction of data-sharing. The aim of this paper is not just to add another reason why it is a good-practice to share data. What we want is to make clear is that sharing data is mandatory in order for science to progress. Each published paper should not be seen as an untouchable revealed truth, but as an opportunity of dialogue, confrontation and verification. This can be done only if all raw data are shared and each researcher has the right tools to reproduce an experiment. Journals like *Cell* and Nature offer unlimited space for fully detailed description of methods. This is very important, because having clear descriptions of the methods used in a study is mandatory, but unfortunately is not enough. In bioscience running exactly the same experiment will not lead to exactly the same data, and therefore researchers cannot verify, for instance, the effect of a different statistical analysis or other post acquisition process applied over the raw data. Only sharing raw data can help science to keep progressing.

Acknowledgements We would like to thanks Prof. Felicitas Kraemer, for her help and support during the early development of the study.

This work was partially supported by the European Community and the autonomous region of Catalonia through the project sMART-O (2013-BP-B-00096).

References

Aggarwal, C. C., & Philip, S. Y. (2008). *A general survey of privacy-preserving data mining models and algorithms* (pp. 11–52). New York: Springer.

Assembly, U. G. (1948). *Universal declaration of human rights.* UN General Assembly.

Barth Jones, D. (2012) *The "re-identification" of governor William Weld's medical information: A critical re-examination of health data identification risks and privacy protections, then and now.*

Begley, G. C., & Ellis, L. M. (2012). Drug development: Raise standards for preclinical cancer research. *Nature, 483*, 531–533.

Bollen, K., Cacioppo, J. T., Kaplan, R., Krosnick, J., & Olds, J. L. (2015) *Social, behavioral, and economic sciences perspectives on robust and reliable science*. Arlington, VA: National Science Foundation.

Bourne, P. E., & Barbour, V. (2011). Ten simple rules for building and maintaining a scientific reputation. *PLOS Computational Biology, 7*(6), e1002108.

Campbell, H. (2014, July 13). The corruption of peer review is harming scientific credibility. *The Wall Street Journal.*

Casacuberta, D., & Estany, A. (2003). *¿Eureka?: el transfondo de un descubrimiento sobre el cáncer y la genética molecular.* Tusquets editores.

Cokol, M., Ozbay, F., & Rodriguez-Esteban, R. (2008). Retraction rates are on the rise. *EMBO Reports, 9*(1), 2.

Collins, F. S., & Tabak, L. A. (2014). NIH plans to enhance reproducibility. *Nature, 505*(7485), 612.

Goodman, S. N., Fanelli, D., & Ioannidis, J. P. (2016). What does research reproducibility mean. *Science Translational Medicine, 8*(341), 341ps12–341ps12.

Hughes, L. D. (2014). The state of science and unreliable research. *Scottish Universities Medical Journal, 3*, s6–s11.

Macherey, P. (2009). Althusser and the concept of the spontaneous philosophy of scientists. *Parrhesia, 6*, 14–27.

Moore, G. E., & Baldwin, T. (1993). *Principia ethica*. Cambridge: Cambridge University Press.

Popper, K. (1963). *Conjectures and refutations: The growth of scientific knowledge*. London: Routledge & Kegan Paul.

Prinz, F., Schlange, T., & Khusru, A. (2011). Believe it or not: How much can we rely on published data on potential drug targets? *Nature Reviews Drug Discovery, 10*, 712.

SINTEF. (2013, May 22). *Big Data – for better or worse. 90% of world's data generated over the last two years*. Last accessed 6/02/2017, from http://www.sintef.no/home/Press-Room/Research-News/Big-Data--for-better-or-worse/

Sokal, A., & Bricmont, J. (1997). *Impostures intellectuelles*. Paris: Odile Jacob.

Sweeney, L. (1997). Weaving technology and policy together to maintain confidentiality. *Journal of Law, Medicine and Ethics, 25*, 98–110.

The Economist. (2013). *Problems with scientific research – How science goes wrong*. Last accessed 6/02/2017, from http://www.economist.com/news/leaders/21588069-scientific-research-has-changed-world-now-it-needs-change-itself-how-science-goes-wrong

The Scientist. (2013). *Fake paper exposes failed peer review*. Last accessed 6/02/2017, from http://www.the-scientist.com/?articles.view/articleNo/37798/title/Fake-Paper-Exposes-Failed-Peer-Review/

Thomson, J. J. (1976). *A defense of abortion. Biomedical ethics and the law* (pp. 39–54). New York: Springer.

US Department of Health and Human Service. (2011). 2011 Office of Research Integrity Annual Report 2011 (US HHS, 2011). Accessed from http://go.nature.com/t7ykc

Vasilevsky, N. A., Brush, M. H., Paddock, H., Ponting, L., Tripathy, S. J., et al. (2013). On the reproducibility of science: Unique identification of research resources in the biomedical literature. *PeerJ, 1*, e148.

Wakefield, A. J., Murch, S. H., Anthony, A., Linnell, J., Casson, D. M., & Malik, M., et al. (2010 [1988]). Ileal-lymphoid-nodular hyperplasia, non-specific colitis, and pervasive developmental disorder in children. *The Lancet, 351*, 637–641 (retracted in Lancet 2010 Feb 2.)

Design Epistemology for Public Health Research

A. Estany

Abstract Design seems to have emerged from the niche in which it had remained for decades, not to say centuries, at least as far as research in the history of philosophy and science are concerned. Nowadays, the idea of design has reached our theories of epistemology: a field that, at first glance, seems to be quite far removed from the analysis of practical situations. The objective of this chapter is to explore just how far design epistemology (DE) can be adopted as a methodological framework for research in the field of public health, and precisely what it adds to classic epistemology or even to the standard methodology of applied science. On the one hand, we will analyse different approaches to DE and related terms and expressions such as "design thinking", "design theory" and "designerly ways of knowing"; on the other, we will consider where to place public health within the field of academic knowledge and research and the disciplines involved in this field of research range from biomedical science to sociology and demography. Finally, we will see how DE can offer proposals and solutions to the challenges that a phenomenon as complex as public health currently faces. That is, we will measure up DE proposals against public health research needs.

Keywords Design epistemology · Design thinking · Cognitive models · Public health

Introduction

There are words that, at a certain time, seem to emerge as catalysts for a series of ideas that had previously been ascribed other interpretations. Among such words are "innovation" and "design." The former was the motivation for a book by A. Estany and Rosa M. Herrera: *Innovation in theoretical and practical knowledge* (2016); the latter is the focus of the philosophical reflection that constitutes this chapter.

A. Estany (✉)
Department of Philosophy, Autonomous University of Barcelona, Barcelona, Spain
e-mail: Anna.estany@uab.cat

J. Vallverdú et al. (eds.), *Philosophical and Methodological Debates in Public Health*, https://doi.org/10.1007/978-3-030-28626-2_7

Design seems to have emerged from the niche in which it had remained for decades, not to say centuries, at least as far as research in the history of philosophy and science are concerned. Even when it has been considered within the academic world, the term has usually been closely linked to art and applied to the different settings of our daily lives. Deciding exactly where the turning point in this subject is located is no easy task, but possibly the industrial revolution, the expansion of manufactured products to broad swathes of the population, mass production and so forth, can be seen as factors that motivated the introduction of design into a vast range of intellectual circles. From the theoretical point of view, the book by H. Simon *The science of the artificial* (1969) represented the introduction of design into the broader academic world, especially as it relates to applied science. In fact, Simon is taken as a point of reference for all things related to design theory, and even today's new up-and-coming authors who aim to go far beyond Simon's original project still hold him in high regard as one of their precursors.

Nowadays, the idea of design has reached our theories of epistemology: a field that, at first glance, seems to be quite far removed from the analysis of practical situations. However, we should bear in mind that epistemology has shifted from an a priori perspective to a naturalised one, in the sense that we cannot engage in epistemology without taking into account the empirical results of science when it comes to configuring methodological models. In addition, the philosophy of science has expanded its field of analysis beyond pure science, and this has made it necessary to consider the epistemology of applied science. In short, design has penetrated into all fields of theoretical and practical knowledge.

The objective of this chapter is to explore just how far design epistemology (DE) can be adopted as a methodological framework for research in the field of public health, and precisely what it adds to classic epistemology or even to the standard methodology of applied science. To this end, we will analyse different approaches to DE and related terms and expressions such as "design thinking", "design theory" and "designerly ways of knowing". One of the issues that we need to address is precisely the polysemy that exists in the field of design, related to many different concepts. Thus, it seems impossible not to engage in a certain amount of conceptual analysis before we can embark on the study of the role of DE in public health research. T. Love (2000) emphasises the lack of clarity in the definition of design concepts in the work of different authors. On this point he states:

> In 1992, I attempted to collect together a glossary of the main theoretical terms of the design research literature, and found that it was an almost impossible task to do justice to the different variants of major terms: there are almost as many different definitions of *design* and *design process* as there are writers about design. (Love 2000: 95)

Faced with this potential confusion concerning the concept of design, Love proposes a study of the structure and dynamics of design theory through meta-theoretical analysis that specifies the different levels of abstraction, ranging from our perception of reality to epistemology and the ontology of design.

Another of the questions that we will consider here is where to place public health within the field of academic knowledge and research. The disciplines involved in this field of research range from biomedical science to sociology and demography. We will examine some of the definitions provided by international organisations and we will locate public health within the framework of healthcare services and their organisation. Finally, we will see how DE can offer proposals and solutions to the challenges that a phenomenon as complex as public health currently faces. That is, we will measure up DE proposals against public health research needs.

Design Epistemology

The field of DE is beyond the classical limits of the branch of philosophy that is epistemology, which seeks to illuminate the foundations of knowledge. The interest that DE has kindled recently responds to several factors, but two of the most important are, on the one hand, naturalist approaches to epistemology and, on the other, the overlapping—though not merging—of scientific knowledge and its practical application. With regard to the former, epistemology has moved forward from its a priori foundation to take into account the empirical results of science. With regard to the latter, the overlap between science and its practical application has meant that it has become necessary for epistemology to move beyond theoretical knowledge, and also to take in its stride practical knowledge and knowhow. Although this overlap is not strictly speaking novel, the speed with which we move from theoretical knowledge to its application, certainly is. The historian of technology M. Kranzberg (1967) makes this point very clearly. He points out that 1700 years went by from the designing of the steam engine in Alexandria until Watt put it to work; it took 300 years for the theory of photograph to be put into practice after it was schematised by Leonardo da Vinci; the electric motor took 40 years to become a practical reality; both nuclear energy and the transistor, just 5 years; transparent plastics, 2 years; and lasers, 18 months! Kranzberg (1968) claims that this progression supports the thesis that the association between science (which aims to discover the "why") and technology (which aims to discover the "how") produces a chain reaction of scientific discovery and technological invention that, although not a new phenomenon, has brought about a qualitative change in the time lapse between the two, which has likewise transformed the characteristics of the relation between science and technology.

DE asks a whole series of questions which, at one and the same time, constitute different perspectives and proposals concerning how to understand the subject of DE itself. On the one hand, we have DE as an alternative to classic epistemology, which is often described as "analytic" and juxtaposed with "synthetic," which is how DE would be described as it would also cover the applied sciences. On the other hand, DE is said to have a series of defining characteristics, among which we can highlight interdisciplinarity as a means of addressing dynamic and complex problems; and a prominent element of social concern expressed through "design thinking" that revolves around human-scale design. Around these principal axes, a series of proposals and considerations arise which we are going to examine in what follows.

To begin with, G. Bengoa (2011) makes a series of observations concerning the scope of science in terms of its practical application. According to Bengoa, the idea that epistemology is concerned with the foundations and methods of scientific knowledge, does not fit with the actual variety of objects that inhabit our field of knowledge; and so there cannot be just one epistemology. In fact, he specifically has in mind constructed objects or artefacts, and he asks whether an epistemology of constructed objects is possible based on parameters other than traditional ones. To this end, he makes a distinction between epistemology "for" design and epistemology "of" design.[1] He says that the former has to do with "a science of knowledge that assists the designer"; while the latter has to do with "an epistemology to use as a tool to examine and learn about the reality of design itself".[2] With respect to the former, we could say that it matches an epistemology of applied science; and in this sense, it would be along the lines of the design methodologies proposed by R. J. McCrory, M. Asimov and A. D. Hall, among others (McCrory 1974). With respect to the latter perspective, the definition does not seem to be very clear at all; unless we interpret it as "an epistemology that uses design itself as a tool to examine and learn about reality". This interpretation could be understood along the lines that the theories of design that we use when constructing objects could be a model for epistemology, both in its function of grounding pure science, in the sense of learning about and explaining reality, and as an epistemology/methodology of applied science. It could be objected here that such a line of argument constitutes a vicious circle; so let us consider it, not as a vicious circle, but as a feedback loop between knowledge, artefacts and design. It must also be made clear that not all authors make such a distinction and that many simply refer to DE. Given the aim of this chapter, focused as it is on the importance of DE for the health sciences, from the distinction that Bengoa makes, we are going to adopt, first and foremost, the sense of epistemology for design. However, we cannot rule out the possibility that the progress of DE has consequences for the epistemology *of* design. That is, although DE focuses on

[1]I wonder whether this distinction, in terms of the definition it gives, would be equivalent to "epistemology of design" (which Bengoa calls "epistemology for design") and "design epistemology" (which Bengoa calls "epistemology of design"). It is a linguistic question but it seems more appropriate to me.

A distinction that could help us to clarify Bengoa's distinction is that between "ethics of neuroscience" and "neuroethics". The former studies the ethical implications of the progress made by neuroscientists and the practices of professionals in this field; while the latter takes neuroscience as a basis for understanding and explaining social, moral and broadly philosophical decisions. We could also bring to mind other such distinctions, like that between "philosophy of cognitive science" and the "cognitive approach to the philosophy of science". The former refers to the philosophical analysis of cognitive science, just as we refer to the philosophy of physics, of chemistry, of biology or of the social sciences; while the latter refers to models of science that are grounded in cognitive science: an approach pioneered by R. Giere, among others.

[2]In the original Spanish, this reads: "una epistemología que utilice como herramienta para conocer a la realidad del propio diseño". The translation is ambiguous insofar as it is not clear whether we are to use design itself to examine reality (see the alternative interpretation later in the paragraph) or, as I have rendered it here, to use epistemology to examine design itself, which sounds rather vague to say the least.

applied science, it also questions some of the criteria of epistemology of pure science. Therefore, analytic epistemology could be brought into question and, consequently, reformulated based on DE as synthetic epistemology.

D. Mahdjoubi (2003), in his paper "Epistemology of design" classifies design as activity, as planning and as epistemology. As activity, the term refers to the conceptualisation (pre-execution) stages of making new products; as planning, to the organisation of the actions that are required to actually bring the product into being; and as epistemology, to the relation with the synthetic methodologies that are required for any change in the applied sciences and distinct from analytic methodologies that are crucial for the progress of any scientific initiative. An important issue that Mahdjoubi raises is the connection between epistemology and methodology. Sometimes, in the philosophy of science, these terms are used indistinctly, and their use may even depend on the tradition one comes from. In general, the French tradition tends to use "epistemology" and the Anglo-Saxon, "methodology". We could further say that epistemology places the emphasis on the criteria for assessing the grounding of knowledge; and methodology, on the processes through which that grounding is performed.

Mahdjoubi points out that analytic methodology has been shown to have shortcomings when addressing applied science, which have allowed synthetic methodology to flourish. Thus, the epistemology of design has been taken as an alternative that can alleviate the deficiencies, especially in fields such as engineering. Once again, this fits in with the proposal of McCrory, Asimov and Hall for a methodology of design that goes beyond the classic methodological schema of pure science (McCrory 1974).

Under the title of "Design Epistemology", D. Karabeg (2012) proposes design as the alternative to tradition. This implies questioning the whole of traditional epistemology which does not sit happily with a move towards innovation as the central axis of scientific research. However, it should be noted that Karabeg is not only thinking of design science, which he does not make any explicit reference to, when he proposes DE; rather he is referring to science in general, both pure and applied. The main idea is what he calls "postulating an epistemology," that is, considering precisely what meaning we are going to give to "epistemology," which goes far beyond the epistemology traditionally understood as the base on which the search for truth and the meaning of things rests.

Therefore, following on from Karabeg, DE reinforces an approach within the framework of academic research that is an alternative to traditional approaches. If we considered the alternative for pure science, it would be what we call "epistemological innovation," which would involve both new forms representation of knowledge and changes in epistemic values, or at least in their prioritisation. In this sense, and being faithful to Karabeg, this epistemological innovation should be performed according to design models. His proposal seems to suggest that, compared to traditional epistemology, that of design is more dynamic and allows new elements that emerge as research progresses to be incorporated into it. That is to say, the introduction of innovation into classic or analytic epistemology based on design models is equivalent to what Bengoa calls "epistemology of design". This is a similar sense to Love's "design philosophy", which he distinguishes from "philosophy of design". The latter, he situates on the same level as the philosophy of technology and of science, that is, the philosophical study of the scientific method. Meanwhile, he considers that the scope of the former is limited to the philosophical study of the method of design. Therefore, it seems that there is a meaning of DE, understood as epistemology of design (Bengoa) or design philosophy (Love), which would involve a naturalisation of epistemology through the science of design. However, as has already been mentioned, given that the subject of the current chapter is the importance of DE in public health research, we are going to adopt the meaning of epistemology for design (Bengoa) or philosophy of design (Love) which, in addition, is the most generalised meaning in design research.

Regarding the characteristics that Karabeg (2012) points out, what he refers to as "wholeness" is particularly important. He defines this as: "the quality that characterizes a perfectly developed and healthy organism, or a complete and immaculately functioning mechanism: All the parts work well and in synergy with one another, and fulfill their purpose within the whole, so that the whole can function well and fulfill its purposes within even larger wholes" (Karabeg 2012: 3). Consequently, we would have transdisciplinarity, which would lead to the Knowledge Federation: a federation of knowledge and disciplines that would allow any phenomenon to be considered and analysed from a whole kaleidoscope of perspectives. This is an important element if we are considering public health research.

Another of the concepts related to DE is "design theory" (DT), which L. E. Östman (2005a, b) writes about in his paper "Design theory is a philosophical discipline—Reframing the epistemological issues in design theory". According to Östman, DT is not a social science or a natural science, but a philosophical enterprise which takes pragmatism as the framework that provides reasons for itself. It is not a case, therefore, of limiting knowledge to a simple and fixed truth; but rather of addressing problems and promoting understanding through clarification, reasoning and criticism. These claims constitute the base of his proposal, focused on knowledge for the resolution of problems, not with a view to discovering truth, or at least not limited to this and not as the priority.

Östman distinguishes between "design research" (DR), whose field of study is design, and "design theory" (DT), which is the stock of knowledge shared by the various fields where design processes take place. He relates DT to Simon, which gives us an indication of his interest in applied science, despite differences in the concept of theory.

> As I see it Design theory matches Simon's "science of the artificial" but the connection to science is not appropriate, whereas the word theory denotes the fact that it is mainly a theoretical discipline in the generally very practice oriented design field. (Östman 2005a, b: 9)

The conclusion that we can draw from this is twofold. On the one hand, DT corresponds to Simon's science of the artificial and, as a consequence, to design science. On the other, however, Östman wishes to emphasise the difference in the meaning that, according to him, Simon wishes to give to the word "theory": a meaning of theoretical discipline that does not fit in with the practical orientation of the field of design. However, it must be said that Simon is extremely well aware of this practical orientation in his work *The science of the artificial.*

Östman also relates DT to "design knowledge" (DK) in the sense that knowledge is an ingredient of both action and research in the field of design. This is an important question that we will deal with next.

Design and Knowledge

Several authors specifically address the relation between design and knowledge; of which, we will concentrate on N. Cross (2006) and A. Rylander (2009). In his book *Designerly ways of knowing* (DWK), Cross considers design as a form of knowledge, intertwined with DE and design thinking. One of the central ideas of his proposal is that design constitutes the third culture, in addition to the two cultures of science and humanities. He therefore thinks that it should form a major part of formal education; not just of training for certain professions, but as one of the fundamental pillars of general education:

> Design in general education is not primarily a preparation for a career, nor is it primarily a training in useful productive skills for 'doing and making' in industry. It must be defined in terms of the intrinsic values of education. (Cross 2006: 5)

Cross points out some differences between these three cultures, with respect to the object of study, the method and the values. In the case of design, the object of study is the artificial world; among the methods, he cites the creation of patterns and synthesis; and the predominant values would be practicality, empathy and matching things to the goals one wishes to achieve.

Other characteristics that are important in DWK include: the manipulation of non-verbal codes in the material culture, the link between doing and thinking, and the importance of iconic modes of cognition (Cross 2006: 11). When it comes to the skills necessary for design, Cross highlights: resolving poorly defined, fuzzy problems; adopting solution-based strategies; and using abductive reasoning together with non-verbal means and graphics to represent knowledge (Cross 2006: 20). He believes that the use of graphics ("sketching") is particularly important as a design tool:

> The use of sketches is clearly an important part of the natural processes of designing, but trying to understand just what this importance is, is something that has only relatively recently become a subject of more careful consideration and analysis by design researchers. (Cross 2006: 34)

Cross also addresses the relation between science and design (Cross 2006: 97), pointing out some differences:

> Scientists try to identify the components of existing structures, designers try to shape the components of new structures. (Alexander 1964)

> The scientific method is a pattern of problem-solving behaviour employed in finding out the nature of what exists, whereas the design method is a pattern of behaviour employed in inventing things of value which do not yet exist. Science is analytic; design is constructive. (Gregory 1966)

Based on this connection, he puts forward three concepts in which we find both the terms “science” and “design”, but expressed in different ways and with different meanings. Thus, he distinguishes between:

(a) Scientific design, which he defines in the following terms:

> Scientific design refers to modern, industrialised design—as distinct from pre-industrial, craft-oriented design—based on scientific knowledge but utilising a mix of both intuitive and non-intuitive design methods.

It would be in order to design, in whatever field although especially in industrial design, that the designer would refer to scientific knowledge.

(b) Design science, which he defines in the following terms:

> Design science refers to an explicitly organised, rational and wholly systematic approach to design; not just the utilisation of scientific knowledge of artefacts, but design in some sense a scientific activity itself.

So, by design science we can understand those sciences whose ultimate goal is not to describe the universe, but to transform it: the different branches of engineering, medicine, computer science, and education science (or pedagogy), etc.

(c) Science of design, which he defines in the following terms:

> The science of design refers to that body of work which attempts to improve our understanding of design through ‘scientific’ (i.e., systematic, reliable) methods of investigation.

Therefore, the science of design is to be understood as the body of theories of design that can be applied in order to bring about a specific product through the scientific method. It should be noted that Cross wishes to make it perfectly clear that the “science of design” is not the same as “design science”.

Cross studies the stages of the design process which, in general terms, can be boiled down to: the formulation of the problem, the generation of the solution, and the strategy to be adopted in the process. All of which constitutes the production of the design project.

We can consider whether the distinctions proposed by Cross are fruitful when it comes to clarifying the complex and wide-ranging field of design. The first thing we need to say is that in practice, that is, in any activity in which design is involved, these concepts are intertwined and they converge in the activity, the product or the process of design. However, precisely because it is a relatively new field, at least in the academic world, an initial conceptual analysis is important. For example, in a design science, let us take medicine for example, the design of radiotherapy apparatus will be a scientific design since it will be executed in accordance with scientific knowledge. In addition, scientific research methods will be followed.

In “Design Thinking as Knowledge Work: Epistemological Foundations and Practical Implications” A. Rylander (2009) explores the different forms of constructing knowledge in the case of business discourse in which the concepts of “knowledge work” and “design thinking” are incorporated. Also in the field of business we encounter two approaches based on two epistemologies: one, rational and analytic;

the other, interpretive and emergent. The former rejects any form of tacit knowledge that cannot be verbalised and validated in an argumentative process; while the latter places the emphasis on designers who learn to "think with their hands" (Collopy 2004), using schemas, prototypes and intuitions to arrive at the ultimate solution (Schön 1983). Rylander (2009: 6) concludes that: "It is argued here that the differing constructions are rooted in the differing epistemological traditions of the fields of knowledge work and design thinking".

Rylander (2009) also points out that just as with "knowledge work" the term "design thinking" is composed of two words that are both rather ambiguous, which makes it difficult to define the composite terms. She attributes the idea that there is no "right" or "wrong" solution, but only "better" or "worse" ones, to Rittel and Webber (1973). Rylander goes on to make a series of comments concerning how to apply design to the world of business, as the following quotations demonstrate:

> Both depend on generating and testing solutions (hypothesis), but the scientific method seeks to uncover what is, while the aim of design is often to envision what might be, but is not yet (Rylander 2009: 5).

> Design firms and knowledge-intensive firms depart from the same basic premise: solving complex problems creatively under ambiguous and uncertain conditions (Rylander 2009: 8).

> By comparing the two different perspectives on problem-solving and highlighting their different epistemological roots, research traditions, and practical implications, the analysis above reveals gaps where the two perspectives could cross-fertilize each other (Rylander 2009: 12).

Although at first glance the world of business may seem very far removed from the aims of this chapter, we must not forget that a large proportion of healthcare institutions function according to business criteria. Therefore, everything that is related to Rylander's reflections on the resolution of problems is relevant for healthcare organisation, which is at the foundation of public health.

Design Thinking

The idea of "design thinking" is central to DE, particularly as synthetic epistemology: an alternative to analytic epistemology. The different meanings of DE share some common characteristics which indicate the importance of the idea of design thinking. One of those shared characteristics that is really at the heart of all of them is the resolution of problems that society is facing and which humans are fully immersed in. Here too we encounter meanings of design thinking that vary, without being at odds with each other, though all of them have to do with human-scale design. For example, T. Lockwood (2009) defines "design thinking" as the process of innovation centred on the human factor that emphasises observation, collaboration and fast learning. It involves applying the sensitivity and methods of the designer to problem solving in different contexts: business, commercial, leadership, services (public and private), etc. In short, the idea of "design thinking" is a way to address the process of innovation focused on meeting the needs of humans in the best possible way. The importance of this approach stems from placing design at the centre of innovation, which is not a move to be taken lightly if we take into account the cognitive capacities of humans.[3]

One of examples Lockwood (2009) offers is that of ski clothing. For this kind of clothing, one of the key goals is protection from the cold, so the materials used are key in achieving this. Successful innovation in this field can be put down to the collaboration of various professionals; here, specifically, designers, engineers and entrepreneurs. It is not necessary for the different inputs to come from different people, but it is necessary for all the factors to converge in the product. In this particular case, according to Lockwood, it was he who contributed the design and commercial part, thanks to his undergraduate business management studies while at university. The engineer was essential to provide knowledge of materials and energy. For the evaluation by end users, on the one hand, the new clothes were tried and tested by regular skiers; and on the other, it just so happened that one of the people who evaluated the commercial possibilities of the ski suits was a major fan of skiing. To sum up, given that any innovation involves different factors that cannot be ignored, a successful novel product can only be the result of the participation of different relevant actors, in this case: engineers, designers and users. Moreover, none of this is of any use if we ignore economic viability based on the costs and benefits for the company behind the project.

Bringing all the factors together is not at all a straightforward affair and it also depends on the predominant trends at any given time. We must not overlook the fact that the industrial revolution eventually resulted in the possibility to mass produce different products, from cars to washing machines, in addition to furniture and buildings. This gave rise to two principal differentiated approaches: one focused on industry and represented by Carnegie, Rockefeller, J. P. Morgan and Ford; and the other which aims to continue the tradition of handicrafts, represented by Charles Rennie Mackintosh, Frank Lloyd Wright and Gustav Stickley (Vogel 2009: 5). C. M. Vogel believes that design thinking can bridge the gap between intensive production, focusing on cost–benefit, and production on a human scale. Vogel (2009: 5) cites the architect Peter Behrens and Bauhaus as examples of design thinking that have attempted to bring together and blend the technological and handicraft approaches. Behrens was contracted in 1907 by Emile Rathenau, the founding president of the German electrical equipment company AEG, to make electricity more acceptable to consumers. Another example of an attempt to bridge the gap is the Bauhaus movement, founded in 1919 by Gropius, whose philosophy was to seek a balance between art, science and mass production. Vogel also portrays design as a way to promote social change and environmental responsibility; it should not only have repercussions for artefacts, but also for business and academic organisation, and so forth. Here we are entering into what is classified as social

[3]In section "The Importance of Cognitive Models" we consider cognitive models, which are particularly important for human-scale design.

innovation and which is well beyond the limits of the business world. Along these lines, Vogel refers to Ralph Caplan, who in his book *By design* (1982) points out that Mahatma Gandhi's concept of non-violent protest constituted one of the most effective designs in history. There can be no doubt that social innovation can play a very important role in public health programmes, via healthcare institutions and organisations.

V. Papanek, in his book *Design for the Real World* (1976), defends design thinking as an approach that focuses on social responsibility in production. He criticises designers who only take into account consumers with a high economic potential. In this way, it becomes important to evaluate user satisfaction with any type of product, taking into account parameters such as usability, accessibility, comprehension and experience. This is one more indicator of the degree of democratisation of innovation.

To summarise, design thinking requires an awareness of the individual person, but also of society as a whole, and of social and political issues that must be in consonance with ethical values in the design process. All of this fits in very well with DE, which goes beyond the search for truth and moves into the terrain of applying scientific knowledge to the resolution of problems facing society.

The Importance of Cognitive Models

We can see that the idea of design thinking implies taking into account cognitive, psychological and social factors, among others, in the process of design. The authors we have referred to here, together with the examples offered, show concern for social responsibility, ethical values, human needs and problem solving; but there is no mention anywhere of the role that cognitive factors can play in achieving the proposed objectives. Neither does McCrory (1974) take into consideration the cognitive factors in his model of methodology of design, even though he introduces scientific and contextual factors, making an important breakthrough for engineering and, in general, for the applied sciences. The question is how to introduce cognitive factors into a design science such as medicine: a discipline that much of public health research is centred on. At this point, cognitive models enter the scene. It is clear that cognitive science cannot be kept separate from technological design. One of the writers who has worked in this field is D. Norman who refers to "cognitive engineering" as cognitive science applied to the design and construction of machines (Norman 1986). Therefore, it is important to include cognitive factors in the case explained by Lockwood, in the approaches analysed by Vogel, in the reflexions of Papanek (1976) and in McCrory's model of methodology of design. Norman (1986) points out the importance of taking into account the fundamental principles that underpin human action with a view to designing systems that require the least possible effort on the part of the user.

One of the first difficulties that cognitive engineering encounters is the difference between psychological and physical variables. People need to control a series of physical systems in order to achieve the goals or objectives that they have set themselves. People who are going to use the machine must interpret the physical variables in order to achieve their goals. To do this, they will need to carry out specific actions and handle specific mechanisms. Any given task, no matter how simple it may be, requires the consideration of various factors and especially the gap between a person's goals, expressed in psychological terms, and the physical system, defined in terms of physical variables.

Norman (2004) puts forward an outline of how the gap between goals and the physical system may be bridged. Each bridge is one-directional, that is, the execution of a task moves from the goals to the physical system, while the evaluation of a task moves in the opposite direction. The first bridge begins with the intentions of the user, then moves to the specification of actions and finally contact is made with the physical system. The second bridge involves interpreting the state of the system by comparing it with the original goals and intentions of the user. It is important to remember that the gap may be bridged either by bringing the system to the user or taking the user to the system. Ultimately, the direction chosen will depend on whether we wish to prioritise the user or the system.

One way of bridging the gap between goals and the physical system is by means of a conceptual model of the system, which acts as a kind of scaffolding on which to build the bridge. The conceptual model, as a mental model, provides us with a very clear idea of how designer, user and system interact. As the outline indicates, the designer builds a conceptual model according to the characteristics of the user and gives it material form in a physical system and from this image users build their conceptual models. In other words, a user's conceptual model is the result of interpretation of the image of the system.

As a result, the designer must try to match the user's model to the design model. In this respect, the designer bares considerable responsibility, given that users constructs their models from the image of the system. The more explicit, intelligible and well-defined the image is, the better able the user is to understand the image of the system. In general, however, the designer must aim to build physical systems that are as explicit and intelligible as possible. Norman (2004) points out that the image of the system takes on even more importance since many people do not read the instructions, although they are incorporated into the system.

Where do we situate these ideas on cognitive engineering in design methodology? Taking McCrory's model, it may be argued that, in the same way as physical and chemical theories (in the case of engineering) depend on the issue to be resolved; and sociological and anthropological ideas depend on needs; so cognitive processes depend on the principles of design, and this constitutes a new discipline which draws on psychology and computer science.

All of the above examples and models are conceived with engineers in mind. Nevertheless, any design principle may be applied to other fields. Indeed, the model that relates designer, user and system could be anything from a tap, plugs, the dashboard of a car or a list of medicine to be taken, to a whole range of products that come from the application of scientific knowledge to the satisfaction of human needs. In the case of public health, we find many elements, from surgical instruments to dietary norms and social welfare programmes, whose design can facilitate or hamper the achieving of levels of welfare, both of individuals and of populations.

Abduction and Pragmatism: Logical-Epistemological Models in Design Theory

One of the issues that DE addresses is the suitability of classic logical-epistemological models for DE. We have already seen that one of the definitions involved presenting DE as an alternative to analytic epistemology, preferring what has been called "synthetic" epistemology. In this vein, we could consider abduction and pragmatism to be models that form the base of such an epistemology.

L. E. Östman (2005a, b) devoted a book to the pragmatist theory of design and its impact on architecture.[4] He considers "the pragmatist philosophy of Dewey as an appropriate philosophical foundation for design theoretical reasoning, as it includes a distinctive theory of knowledge and research. It can integrate both the scientific methods, the reflective and interpretative methods of the social sciences and the humanities, but it cannot be fully understood in the light of one of these. It needs its own paradigm or epistemology" (Östman 2005a, b: 11). One of the elements that he considers to be particularly interesting for DE is that "Dewey's philosophy includes a great deal of ethical convictions, which he saw as grounded in logical reasoning rather than moral imperatives" (Östman 2005a, b: 12). Among its virtues we find the following: "The pragmatist philosophy of Dewey basically provides a theoretical model of creative design actions, with experimentalism and evaluative steps as the core elements—but due to a capability that one has to learn by training in an adequate design community" (Östman 2005a, b: 17). This represents an entire series of characteristics that pertain to reasoning in biomedical science.

The pragmatic perspective is behind the majority of approaches to design theory, insofar as they provide lenses through which concepts can be interpreted, as well as a set of conceptual tools that support the explanation, understanding and communication of what design is. According to Östman, this perspective would become evident through a series of aspects such as: design being a controlled creative action; design knowledge being a set of repertoires that attempt to address problems, desires, values and complex situations, and the objective of which is to change an existing situation into a preferred one; a design not being a single simple issue, but rather a complex situation where we want to improve or change something; and creativity being a dynamic capacity to manage problems and situations, and to generate solutions that match expectations (Östman 2005a, b: 8). Based on these considerations,

[4] Although Östman applies it to architecture, the general principles and the criteria for considering DE to be suitable for architecture are the same as for its suitability for medicine and overall for research in health sciences.

pragmatism becomes the philosophical substrate of DE. Positivism and logical empiricism are left far behind as the epistemological underpinning of science.

P. Dalsgaard (2014) also backs pragmatism, and considers that it constitutes conceptual scaffolding for "design thinking". As he tells is: "I employ the term design thinking in an inclusive sense to denote the ways of framing, approaching and addressing challenges that characterize design" (Dalsgaard 2014: 144). The pragmatic perspective can be of great value in terms of at least four aspects: the theory/practical relationship, the interrelation between experimentation and technology, the implications of a pragmatic perspective for design and the application of the concepts of pragmatism in different fields such as art and education. These four aspects are not the only factors that intervene in design, but there is no doubt that they constitute key issues for its study and debate. Dalsgaard (2014) concludes: "The main argument presented in this article is that Deweyan pragmatism offers a set of concepts that can contribute to the efforts to articulate designerly inquiry and thinking, as well as a framework for understanding the relations between these concepts" (Dalsgaard 2014: 52).

K. Dorst (2011) asks: "What is the core of Design Thinking?" and "What could it bring to practitioners and organisations in other fields?" Centring his enquiry on these questions, he considers that logic provides us with a set of concepts that demonstrate the form of reasoning in design to be a new paradigm that we can use to tackle problems in different professions, particularly in IT (e.g., Brooks 2010) and business (e.g., Martin 2009). The basic pattern of reasoning would be abduction, that is, abduction as a logical model of design, and pragmatism a meta-epistemology of design.

In "Abductive Thinking and Sensemaking: The Drivers of Design Synthesis", J. Kolko (2010) supports abductive thought for what he calls "design synthesis." In this case, he uses the idea of synthesis not as a synthetic epistemology that is an alternative to analytic epistemology, but as the way to design based on summarising all the elements that the designer has managed to obtain for the elaboration of the artefact in question. By way of conclusion he says:

> This paper has defined design synthesis as an abductive sensemaking process of manipulating, organizing, pruning, and filtering data in the context of a design problem, in an effort to produce information and knowledge, and has introduced three methods of formalizing the synthesis process in practice. Each of the methods—reframing, concept mapping, and insight combination—emphasizes prioritizing, judging, and forging connections. These qualities are derived directly from the logical processes of abduction and the cognitive psychology theory of sensemaking. (Kolko 2010: 26)

In many of the analyses of design and its relation with abduction and pragmatism, we see rekindled the debate between an epistemology grounded in the linguistic turn (verbalisation of knowledge) and current approaches in which linguistic representation is not essential, since there are other means of representing knowledge, including schemas, gestures, etc. Although it is not usually referred to directly, the alternative that focuses on design borrows characteristics from the cognitive approach in epistemology and philosophy of science. We will not go into this matter in detail here, but a study of the role of cognitive models in many of the proposals regarding design theory is overdue.

Public Health Research

Public health is a complex and important filed for any society. This complexity requires the intervention of many actors when dealing with it, and for this reason research also needs to incorporate knowledge from different disciplines and to reconsider the classic methodological models of science. One of the first issues to be addressed is its definition, since the term implies the broadest possible scope. In the book *Public health: ethical issues* (2007), published by the Nuffield Council on Bioethics, we find a series of definitions that lay out the general lines of what is understood by public health.

> The science and art of preventing disease, prolonging life and promoting health through organised efforts of society. *Faculty of Public Health of the Royal Colleges of Physicians of the United Kingdom*
>
> The mission of public health is to 'fulfil society's interest in assuring conditions in which people can be healthy'. *Institute of Medicine, USA*
>
> Public Health is the science of protecting and improving the health of communities through education, promotion of healthy lifestyles, and research for disease and injury prevention *Association of Schools of Public Health, USA*
>
> An art and a science; and also a movement dedicated to the equitable improvement in health and well-being (of communities with their full participation). *World Federation of Public Health Associations*
>
> The key functions of public health agencies are assessing community health needs and marshalling the resources for responding to them, developing health policy in response to specific community and national health needs, and assuring that conditions contributing to good health, including high-quality medical services, safe water supplies, good nutrition, unpolluted atmospheres and environments that offer opportunities for exercise and recreation are available to the individuals. *World Medical Association*
>
> The science and the art of (1) preventing disease, (2) prolonging life, and (3) organized community efforts for (a) the sanitation of the environment, (b) the control of communicable infections, (c) the education of the individual in personal hygiene, (d) the organization of medical and nursing services for the early diagnosis and preventive treatment of disease, and (e) the development of the social machinery to ensure everyone a standard of living adequate for the maintenance of health, so organizing these benefits as to enable every citizen to realize his birthright of health and longevity. *CEA Winslow, former Chair of Department and Professor of Public Health, Yale University*. (Box 1.4: Public health: definitions, p. 6)

Based on these definitions, we can indicate some of the characteristics that, in one way or another, they all recognise: the impact healthcare has on society; the aim of ensuring a good quality of life for the members of society; the development of healthcare policies that promote medical services as a response to the needs of individuals; and the prevention of diseases through social initiatives that range from education to prevention through diagnosis in populations particularly at risk of specific diseases.

Delving a little deeper into the conception of public health, one important reference is the book *The future of public health* by the "Committee for the Study of the Future of Public Health" and published by the National Academy of Sciences, in 1988. The Committee's aim is to demonstrate the difficulties faced by public health in the United States, in order to stimulate actions aimed at ameliorating the health problems that the population suffers. This requires addressing, on one hand, the scientific and technical level; and, on the other, ethical values and governmental responsibility. Thus, the Committee defines public health in the following terms: "the organized community efforts aimed at the prevention of disease and promotion of health. It links many disciplines and rests upon the scientific core of epidemiology" (p. 41). And the World Health Organization (WHO) defines "health" as "a state of complete well-being, physical, social and mental, and not merely the absence of disease or infirmity."

Given the challenges that public health faces, one of the most important questions is which research model will be the best suited to this field in which knowledge of several disciplines pays an important role. J. Frenk, J. L. Bobadilla, J. Sepúlveda, J. Rosenthal and E. Ruelas (1986) pose a highly pertinent series of questions in this respect: What is the field of action of public health research? How does it relate to health service research and socio-medical research? In what way does it incorporate the methods and theories of social and biomedical sciences? Does it have to be oriented towards the solution of practical problems or towards the advance of knowledge and the integration of a coherent body of findings and principles? The authors make some proposals that can be very useful in answering these questions.

They propose a typology of health research,[5] beginning with a certain level of analysis (individual and population) and of the subject of the analysis, distinguishing between conditions and responses. The final result is that at the individual level, we have biomedical research that provides answers to clinical research. At the level of population, we have research into health needs, the objective of which is to provide answers for research into systems of health. With this proposal, we can see that, on the one hand, we have pure or descriptive science in the form of biomedical science, and the applied science of clinical research; while on the other, we have health needs, for which answers are sought in the different systems of healthcare organisation, which are mixed together with social and policy factors, in short, contextual factors.

This typology is accompanied by some research modules relevant for public health, which we can see as dimensions of the research in public health. They offer an overview of all the elements that have an impact on research, from

[5]The literature on research in health public is vast and it would be impossible to give an exhaustive overview of it in this work. The choice of the proposal by Frenk et al. in no way means that it is the only adequate one for this research but we can say that it contains the elements necessary to undertake it, it integrates the factors that intervene in public health and it provides a proposal for an interdisciplinary approach.

methodological areas, knowledge areas and substantive areas (the phenomena that are the objects of public health research). The methodological areas include data analysis, implementation and design. The different areas of knowledge form a series of disciplines that are involved in public health such as epidemiology, demography, economy, sociology and administration sciences. Meanwhile, in the substantive areas we find services utilisation, resource availability and the intervening factors between needs and services, and between services and resources. The resolving of problems is implicit to these dimensions, and depending on the specific characteristics of each problem, we would address with recourse to one or other discipline, method and form of organisation. We could say that public health research does not contemplate knowledge for its own sake, but as a conceptual instrument to help us meet the challenges that we face.

It should be said, however, that neither the typology nor the interdisciplinary dimensions can be interpreted as consisting of isolated compartments. That would not fit in at all with the interdisciplinary vision that underlies the entire proposal, at least implicitly.

From this proposal we can draw several conclusions regarding public health research: there is a need for an interdisciplinary approach; from the methodological point of view, we must take into account both the classic models of pure science and those of applied science; public authorities must be involved in order to achieve the proposed goals; and everything must be viewed from the perspective of moral and ethical values. These conclusions are in no way intended to offer an exhaustive overview of all public health research, but rather to point out some of the issues that constitute the common denominator of any project involving public health research.

The Importance of Design Epistemology for Public Health Research

Based on the analysis of DE and considerations of what we understand public health to be, we are going to indicate some elements that are particularly relevant when it comes to bringing the two together. We will focus on the models that are at the base of that research and constitute its foundations, paying attention to the factors that come into play in the broad and complex field of public health.

Just by contemplating the definitions that we have presented, we can see that both research in this area and the possible solutions to problems require the convergence of different factors, of a range of disciplines and of both cognitive and contextual values. All of this takes place within the context of the interrelation between theoretical and practical knowledge; both of which are necessary but neither sufficient on its own. From this we can say that the characteristics of public health research encounter in DE proposals that address the challenges it faces.

As an initial consideration, and taking into account the fact that biomedical science, which we have qualified as pure or descriptive science, is involved in public health research, there is a part of analytic epistemology that will also necessarily play a role in the research, even though it is actually brought into question by DE. However, public health is centred on applied science and the resolution of problems, and it is here that the exclusively analytic epistemology proves to be insufficient. Therefore, in analysing the importance of DE for public health research, we will focus on the part that pertains to applied science, and analyse some of the elements of its foundations.

Theoretical Framework

The theoretical framework that inspires applied science is what H. Simon calls "science of the artificial" and also "design science", defined as science that does not only describe the world, but rather it aims to change it. Within this broad classification, we find fields ranging from engineering to education science (pedagogy). The philosopher I. Niiniluoto (1993) uses Simon's idea as a reference for the philosophy of design science.

For the case at hand, there are several different design sciences that make up the field of public health, such as medicine and epidemiology, principally. This certainly does not imply that disciplines such as biology, chemistry and what has come to be called "biomedical sciences" do not constitute the theoretical support that allows progress to be made in the resolution of the problems faced by public health. DE would also form part of the theoretical framework, moving beyond the search for truth and, in the case of public health, its inescapable objective would be the resolution of practical problems.

Methodology

Public health research methodology cannot be the standard method but rather it must be one of the methodological models of design, such as those of McCrory, Hall, Asimov, etc. (McCrory 1974). In all of them, at the design stage, whether we are dealing with a drug, an aeroplane or a syllabus, scientific knowledge is brought together with needs, and also with a whole range of geopolitical, social and economic factors. In the case of public health, the scientific input would consist of the biomedical sciences and the contextual input would be healthcare organisation, in addition to the data that sociology, demography, etc. could all contribute. Therefore, we can say that the design methodology proposed by McCrory, Hall and Asimov, among others, is suitable for addressing public health research (McCrory 1974).

Philosophical System and Logical Structure

As we show in section "Abduction and Pragmatism: Logical-Epistemological Models in Design Theory", DE has adopted pragmatism as a suitable philosophical approach to form the foundations of design reasoning. Pragmatism places human action at the centre of theoretical reflection, in order to understand the world. Inspired by J. Dewey (1859–1952), Östman indicates that truth is not something that is eternal, but rather the result of continued efforts to understand a specific phenomenon. In short, pragmatist philosophy provides a model for the design of creative actions. In education, the maxim "learning by doing" is heartily defended and, from an ethical point of view, it promotes control of one's actions in an uncertain world.

In this way, given the practical character of public health research and the characteristics of pragmatism, there is no doubt that the two fit together neatly. Abduction, as a form of logical inference, constitutes the logical underpinning of pragmatism.

Cognitive Models

Paying attention to the fact that design thinking is targeted at human-scale design, cognitive models are of primary importance for the designer. Here, the work of D. Norman is an essential reference (Norman 1986). Although Norman takes engineering as his starting point, he applies the cognitive models presented in section "The Importance of Cognitive Models" to any field and process of design. In the case of public health, the list of products in which a transparent or an obscure design could benefit or prejudice the health of a person or a collective would be unending. These would include, by way of example, radiotherapy apparatus, surgical instruments and pill dispensers, but also the design of hospitals, of the healthcare administration and of the systems used to sterilise equipment. Whether we take the cognitive capacities of people, and especially of the users, into account or not, can make the difference between success and failure in public health programmes; and in extreme cases, this could lead to accidents caused by human error.

Indicators of Design Epistemology

DE is posited as an alternative to analytic epistemology, which is insufficient for fields such as public health. Some of the characteristics of DE constitute extremely important elements for public health research. We wish to highlight:

- "Wholeness", in the sense of the integration of different elements within a specific phenomenon.
- Interdisciplinarity, which is vital for such a complex field as public health.

- The need for a "knowledge federation": an idea of Karaberg's.

One element that is appropriate for public health is the perspective of seeing design as a third culture (N. Cross), between science and humanities. In this way it would represent a bridge between those two major fields of knowledge that have provoked much discussion and have appeared irreconcilable, as if they constituted two different worlds. Design addresses issues which necessarily require an overlapping of these two worlds. Cross's proposal is also relevant for the professional aspects of DE; for practically all types of professions, but particularly in education, if we consider design as the third culture.

Professionals of myriad different natures intervene in public health, from doctors and nurses, to healthcare educators, hospital administrators, social workers, etc. Therefore, an epistemological model that takes the different professions into account may be extremely fruitful for the development of healthcare policies, and in this way offer solutions to the challenges that are faced in healthcare.

Finally, it is clear that public health implies activity that requires rigorous planning in order to achieve its objectives. This fits in with Mahdjoubi's idea that classifies design as an activity, as planning and as epistemology.

Conclusions

Public health constitutes one of the most important challenges our society faces, given that a large proportion of the welfare of humanity is at stake in it. On one hand, the consequences for people are huge; and, on the other, the factors that intervene, and on which public health depends, are extremely wide ranging, and therefore so are the disciplines involved in public health research. The most immediate consequence of this is the need for an interdisciplinary approach, both at the level of knowledge and at that of methods.

Classic epistemology has focused on what has been called "pure science". However, as we have seen throughout this chapter, public health research is grounded in both biomedical science and knowledge of the context in which society develops. Therefore, what is also needed is the implication of sociology, demography, politics and ethics. Based on this, we can conclude that applied science is crucial when it comes to addressing and providing solutions for the problems that the field of public health faces.

We need to take into account, on the one hand, the particular characteristics of public health and the range of perspectives from which it can be approached; and on the other, the shortcomings of classic or analytic epistemology. When we do this, we see that design epistemology and everything in its immediate orbit, such as design thinking, design theory, designerly ways of knowing, etc., constitute a suitable framework from which to address public health research. Such an approach can provide methodological instruments that offer solutions to the challenges that society faces in this field. For their part, cognitive models of human capacities provide the elements necessary for design to be transparent rather than obscure, thus facilitating adaptation to people's needs.

Acknowledgements This work was supported by the Spanish Government's Directorate-General for Science Research and Development via research project: FFI2017-85711-P Epistemic Innovation: The Biomedical Sciences Case.

This work is part of the consolidated research network "Grup d'Estudis Humanístics de Ciència I Tecnologia" (GEHUCT) ("Humanistic Studies of Science and Technology Research Group"), recognised and funded by the *Generalitat de Catalunya*, reference 2017 SGR 568.

References

Alexander, C. (1964). *Notes on the synthesis of form*. Cambridge, MA: Harvard University Press.

Bengoa, G. (2011). *Distintos acercamientos epistemológicos: cinco enfoques sobre los objetos*. Seminario en la Facultad de Arquitectura, Diseño y Urbanismo | UBA Universidad de Buenos Aires.

Brooks, F. P. (2010). *The design of design: essays from a computer scientist*. New York: Addison-Wesley Professional.

Caplan, R. (1982). *Why there are no locks on the bathroom doors in the Hotel Louis XIV and other object lesson*. Mcmillan: St Martins Press.

Collopy, F. (2004). "I think with my hands": On balancing the analytical and the intuitive in designing. In R. J. Boland & F. Callopy (Eds.), *Managing as designing*. Stanford, CA: Stanford University Press.

Committee for the Study of the Future of Public Health. (1988). The future of public health. Published by the National Academy of Sciences.

Cross, N. (2006). *Designerly ways of knowing*. London: Springer.

Dalsgaard, P. (2014). Pragmatism and design thinking. *International Journal of Design, 8*(1), 143–155.

Dorst, K. (2011). The core of 'design thinking' and its application. *Design Studies, 32*(6), 521–532.

Estany, A., & Herrera, R. M. (2016). *Innovación en saber teórico y práctico*. London: College Publications.

Frenk, J., Bobadilla, J. L., Sepúlveda, J., Rosenthal, J., & Ruelas, E. (1986). Un modelo conceptual para la investigación en salud pública. *Boletín de la Oficina Sanitaria Panamericana, 101*(5), 477–492.

Gregory, S. A. (1966). Design and the design method. In S. A. Gregory (Ed.), *The design method*. London: Butterworth.

Karabeg, D. (2012). Design Epistemology. *Information, 3*, 1–x.

Kolko, J. (2010). Abductive thinking and sensemaking: The drivers of design synthesis. *MIT's Design Issues, 26*, 1.

Kranzberg, M. (1967). The unity of science-technology. *American Scientist, 55*(1), 48–66.

Kranzberg, M. (1968). The disunity of science-technology. *American Scientist, 56*(1), 21–34.

Lockwood, T. (Ed.). (2009). *Design thinking. Integrating innovation, customer experience, and brand value*. New York: Allworth Press.

Love, T. (2000). Philosophy of design: A metatheoretical structure for design theory. *Design Studies, 21*, 293–313.

Mahdjoubi, D. (2003). Epistemology of design. In *Integrated design and process technology* (pp. 1–5). IDPT, June.

Martin, R. (2009). *The design of business*. Cambridge, MA: Harvard Business Press.

McCrory, R. J. (1974). The design method-A scientific approach to valid design. In F. Rapp (Ed.), *Contributions to a philosophy of technology* (pp. 158–173). Dordrecht: D. Reidel.

Niiniluoto, I. (1993). The aim and structure of applied research. *Erkenntnis, 38*, 1–21.
Norman, D. A. (1986). Cognitive engineering. In D. A. Norman & S. W. Draper (Eds.), *User centered system design. New perspectives on human-computer interaction*. Hillsdale, NJ: Erlbaum.
Norman, D. (2004). Emotional design. Why we love (or hate) everyday things, Basic Books. Versión castellana de: *El diseño emocional. Por qué nos gustan (o no) los objetos cotidianos*, Paidós Ibérica, Barcelona, 2005.
Nuffield Council on Bioethics. (2007). *Public health: ethical issues.*
Östman, L. E. (2005a) *A pragmatist theory of design. The impact of the pragmatist philosophy of John Dewey on architecture and design*. PhD Dissertation School of Architecture Royal Institute of Technology, Stockholm.
Östman, L. E. (2005b). Design theory is a philosophical discipline-Reframing the epistemological issues in design theory. *Design System Evolution.*
Papanek, V. (1976). *Design for the real world: Human ecology and social change*. Frogmore: Paladin.
Rittel, H. W., & Webber, M. M. (1973). Dilemmas in a general theory of planning. *Policy Planning, 4*, 155–169.
Rylander, A. (2009). Design thinking as knowledge work: Epistemological foundations and practical implications. *Design Management Journal, 4*, 7–19.
Schön, D. A. (1983). *The reflective practitioner: How professionals think in action*. New York: Basic Books.
Simon, H. (1969). *The science of the artificial* (1st ed.). Cambridge, MA: MIT (1996, 3rd ed.).
Vogel, C. M. (2009). Notes on the evolution of design thinking: A work in progress. En T. Lockwood (Ed.), *Design thinking. Integrating innovation, customer experience, and brand value*. New York: Allworth Press.

Invisible Victims and Public Health: Epistemic Injustices in the Intergenerational Transmission of Trauma

María del Mar Cabezas and Carlos Pitillas

Abstract This chapter applies the concept of hermeneutical injustice to the social and clinical misconstructions regarding the case of intergenerational transmission of trauma and subtle relational trauma, as well as to the lack of a child centered perspective, obscuring so not only the social study and understanding of the problem, but the victim's skills to make sense of her own experiences from generation to generation.

Mechanisms underlying the intergenerational transmission of trauma, as well as the transformations of trauma across generations, are outlined. Traumatized children may adopt relational strategies that differ in form from the harmful treatment they have received from caregivers. This may translate into behavioral patterns that seem unrelated to the trauma experienced by the child within her attachment relationships, and thus compromise the ability to recognize, prevent and treat that harm effectively, becoming so a problem of justice and public mental health.

In order to overcome the invisibility involved in these cases and its negative impact on public health policies, the chapter defends that those cases constitute an instance of epistemic injustice. Several normative problems and paradoxes concerning the ethics of child health, such as the questions of thresholds in public health policies or the conflicts on parental responsibility, are addressed through the lens of epistemic injustice. As a result, we advocate, contrary to reactive views, for inclusive preventive measures as a matter of justice to avoid the perpetuation of invisible injustices.

Keywords Epistemic injustice · Recognition · Parental responsibility · Attachment · Intergenerational trauma

---

M. del Mar Cabezas (✉)
Facultad de Filosofía, Universidad Complutense de Madrid, Madrid, Spain

C. Pitillas
Facultad de Ciencias Humanas y Sociales, Departamento de Psicología, Universidad Pontificia de Comillas, Madrid, Spain
e-mail: cpitillas@comillas.edu

J. Vallverdú et al. (eds.), *Philosophical and Methodological Debates in Public Health*, https://doi.org/10.1007/978-3-030-28626-2_8

Justice, Public Health, and the Moral Expectation of Recognition

As Jean Améry remarks, a human world is a world where one expects not to be harmed and, if harmed, one expects then to be helped and supported (2001, pp. 91–92). These two principles refer to the idea of moral reciprocity, trust and recognition, which may be threatened every time one faces a problem of justice. As Josep Corbí recently developed, knowing that the others do not want or cannot listen to your story destroys the second expectation (2005, p. 51).

To put it differently, if someone harms you, this implies a lack of recognition of your moral status, whilst the lack of help from the spectator indicates a lack of moral sensibility, a kind of invisibility in the victim or, in other words, a lack of recognition. This lack of recognition executed not only by the perpetrator, but by the moral spectators, is implicitly directed to both, the victim as a valuable being, and the event as a morally relevant one. As such, these two expectations exceptionally captured by Améry denote two problems that may appear every time we face a question of justice, namely, epistemic and concretely hermeneutical injustices (Fricker 2007).

Every time two or more subjects interact in an asymmetrical relationship where one has power over the other, we can affirm that this scenario is susceptible to generating a problem of injustice. Nevertheless, there are other subtle aspects in these interactions that arise from that asymmetrical situation, i.e., not knowing what happens to you, being thought of *not being able to know* and determine what the accurate relevance of what you experience is, not being able to *interpret* the events, as well as lacking any kind of recognition from the community. All these subtle sides may surely have a severe impact on that person in terms of wellbeing, and, from a collective point of view, also on that community in terms of social justice. In a nutshell, not recognizing a damage as such is a way of denying that something morally relevant happened, which, in turn, may become an instance of epistemic injustice, as that society will be tacitly denying the ability of the victim to *know* what is wrong and to evaluate whether that was a damage or not.

As agents and spectators, it might the case that whether we do not see that we are causing a damage (breaking so Améry's first expectation), or, when the impact on the victims' life is ignored or denied, we are not giving the necessary support to the victims (breaking in this case Améry's second basic expectation). At a collective level, these instances of injustice may lead to an invisible type of inequality and discriminations, perpetuating so an unfair scenario.

Moreover, this kind of injustices are more likely to occur (and surely not sufficiently analyzed from a philosophical point of view) when two conditions co-occur, namely, when the agents involved are children, and when mental health and mental well-being is affected more that any other aspect of the person's welfare.

On the one hand, children are an especially vulnerable segment of the population, due to their singular traits and, among many other cultural and developmental reasons, due to their lack of public voice. Likewise, "children are vulnerable because they need care not only to survive but also to develop their basic physical, intellectual, and emotional capacities" (Mullin 2014, p. 266). As a result, children face several types of vulnerability that increase the chances of suffering multiple disadvantages, which, in turn, may become a corrosive social handicap and an injustice. In this sense, although every human being is damageable and interdependent, children are intrinsically dependent on caregivers and, therefore, profoundly vulnerable.

Thus, children are especially prone to experience some sort of epistemic injustice, something that should not be ignored by public health debates, as we will develop in the following sections.

On the other hand, whilst a basic agreement on the relevance of physical intrafamily abuse for children's wellbeing has been achieved, the debate on children's mental health, especially when it comes to subtle traumas, does not find the same social and clinical recognition. There is still a lack of awareness concerning the importance of mental health regarding the debates on public health and social justice, so that common positions swing between the tendency to pathologize normal processes and the tendency to normalize subtle and recurrent damages. Taking this into consideration, epistemic and hermeneutical injustices are more likely to occur when it comes to damages that are subtle or without physical impact, which tend to go unnoticed.

For these reasons, in this chapter we will examine the ethical questions involved in the epistemic injustices related to the intergenerational transmission of trauma, as an under-explored issue that connects the two previously mentioned factors: an especially vulnerable and not always socially and politically visible group of population, and a significant dimension of health for the manifestation of epistemic injustices.

Although other questions related to children's mental health may indeed also be affected by epistemic injustices and may also deserve attention, the challenges related to the continuity/discontinuity of intergenerational trauma transmission imply a set of open problems for social epidemiology and therefore deserve to be addressed from a philosophical perspective as it is a clear example of a problem of public health where epistemic injustice is perpetuated. By delving into the ethical questions concerning children's mental health, we hope to enrich the current debates on social epidemiology.

In order to do so, we will firstly explore the concepts of epistemic and hermeneutical injustice, following Fricker's work, as a potential useful tool to shed some light on some less visible problems of justice affecting children from generation to generation. Secondly, we will begin by considering how much is known about the issue and will examine how early relational trauma may shift in form across generations, impacting on children's development in complex ways. In doing so, we will be able to defend that those cases constitute a case of epistemic injustice that should be taken into consideration in a different way by public health. Finally, we will discuss some paradoxes and normative challenges that the previous considerations may imply for social epidemiology and public health approaches as well as we will raise some questions concerning agents of justice and responsibilities.

Child Vulnerability and Epistemic Injustice

In this section we will explore the concept of *epistemic injustice*. Following Miranda Fricker's work, an epistemic injustice is "a wrong done to someone specifically in their capacity as a knower" (2007, p. 12). She also introduces the distinction between *testimonial injustice* and *hermeneutical injustice*:

> Testimonial injustice occurs when prejudice causes a hearer to give a deflated level of credibility to a speaker's word; hermeneutical injustice occurs at a prior stage, when a gap in collective interpretive resources puts someone at an unfair disadvantage when it comes to making sense of their social experiences. An example of the first might be that the police do not believe you because you are black; an example of the second might be that you suffer sexual harassment in a culture that still lacks that critical concept (2007, p. 12).

Both testimonial and hermeneutical injustices, could be applied to children when their testimonies are not believed because they are children (and "children tend to lie", for example), or when their experiences are "obscured", denied, or neglected (because "it is not so serious", or by any other current social myth or prejudice on the issue). In fact, the prejudice against children's claims, also known as the 'age bias', as well as the prejudices related to mental health and the impact of subtle traumas are so inculcated that, as Bufacchi points out in general terms, they may "fail to shock us" (2016, p. 6), becoming invisible.

Thus, either the victim is not recognized as a valuable being who deserves consideration, or the victim is somehow considered as relevant, but the harm done is not. However, this clear-cut distinction can also be questioned, as these two types of injustices are often deeply interwoven. In many cases the fact that the damage is ignored is also a sign of a lack of social power, recognition and credibility from the part of the victim.

That being said, and although children may frequently be affected by these two types of epistemic injustices, we will pay especial attention to the second type, that is, hermeneutical injustice, as it plays a key role in understanding the effects of discontinuity of intergenerational transmission of trauma and the social attitudes associated with it.

Explicitly defined as "the injustice of having some significant area of one's social experience obscured from collective understanding owing to hermeneutical marginalization[1]" (Fricker 2007, p. 158), hermeneutical injustices may also play a crucial role in social justice and social epidemiology.

In light of Fricker's work, hermeneutical injustice refers therefore to the lack of social recognition and understanding of an unfair event as such, which, in turn, implies deeply negative consequences for the victim. The victim is not only harmed by the experience, but also by the unintelligibility resulting from the lack of collective understanding and recognition. This leads to the breakage of the sense

[1]Miranda Fricker understands hermeneutical marginalization as the unequal participation of the members of a marginalized group regarding relevant aspects of their social experiences (2007, pp. 152–154).

of reciprocity and trust (in oneself and in others), which are basic requirements to properly develop a human life.

The lack of a suitable category also plays a central role. A society that lacks a "word" with a shared and accepted meaning in order to categorize an event or an experience as an instance of injustice, jeopardizes the victim's ability to be aware of the fact that she is indeed a victim of something unfair, and her ability to call for help.

Considering now the intergenerational transmission of trauma and children's early attachment relationships with caregivers, the prevalence of this type of injustice may be due to a lack of a shared concept of child psychological maltreatment or to a dissonance between official, formal—whether legal or clinical specialists'—definitions, and socially extended ones, as Haslanger (2012) points out when she makes the distinction between *manifest concepts* and *operative concepts*, respectively. In other words, although child psychological maltreatment is officially recognized as a relevant problem of public health and a corrosive disadvantage for children's wellbeing (*manifest concept*), the current usage (operative concept), the myths, and prejudices denote that the manifest concept has not been socially integrated yet. As a result, the manifest concept has not turned out to be an operative one.

Analogously to what Katharine Jenkins (2016) defends in relation to women's sexual abuse, it can be argued that children who are traumatized in a different way as expected or by socially invisible or unexpected triggers, suffer a second order injustice when they find difficulties to categorize and communicate their harmful experiences, due to a lack of accurate and socially accepted concepts that reflect the experience and the damage related to it.

As a result, and as we will try to examine in the following sections, some invisible forms of intergenerational transmissions and some social attitudes may constitute a clear case of an epistemic injustice that should receive more attention from a philosophical perspective.

Invisibility of Harm and Invisibility of the Mechanisms Leading to Harm: The Case of Intergenerational Transmission of Trauma

One of the most generalized assumptions about early relational trauma is that it tends to be transmitted to the next generation in a *continuous* or *isomorphic* way. For example, a child who was abused by her parents will be seen as prone to develop forms of parenting that are similar to the ones used upon her, and thus becomes an abusive parent. Continuity in the intergenerational transmission of trauma is not only a general assumption within folk psychology, but has also been partially confirmed through empirical research (Berlin et al. 2011; Pears and Capaldi 2001; Widom et al. 2015). In face of this, we will present arguments and evidence that speak of the presence of *discontinuous* or *heteromorphic* transmission of trauma across generations. Some of the children who were hurt by their parents, may grow into adulthood and become parents who use child-rearing practices that differ from the ones they were exposed to, but which nonetheless may be equally traumatic for their children. These new forms of child traumatization, besides being different from those suffered by the parents in the past, may have a subtle or "silent" quality. This leads to a set of problems of a clinical and ethical nature compromising public health and social justice.

From birth, the child is motivated to create and maintain attachment relationships with caregivers, who can provide two experiences that are essential to survival: protection from danger, and comfort with respect to unregulated emotional states (Bowlby 1969). Most caregivers (normally, the parents) are able to detect and interpret the child's affective cues, provide with adequate responses and, when possible, stimulate the child's exploration (Powell et al. 2013). This set of responses increases the child's probability of survival, and helps her stay within a state of optimal neurophysiological activation (Schore 2010), something that is essential for growing and learning. The result is usually an experience of security in the child. Secure attachment in infancy and childhood leads to a set of socio-emotional skills that, with development, may have a great influence upon the adult's capacity to parent a child. These skills involve, among others: the ability to communicate in an effective, regulated manner, negative emotions (e.g. distress, fear, anger, need for comfort); a sense of trust in adult attachment figures (such as spouses or close friends); the ability to ask for help; adequate strategies of emotion regulation; and a balanced (i.e. not prone to distortions) processing of social information. These skills, acquired within the earliest relationships, help parents express their difficulties, lean on others for help, self-regulate in face of parenting-related stress, maintain a balanced view of themselves as caregivers and of their child, or apply parenting practices with flexibility, among others.

However, attachment security is inaccessible for some children. These children may be exposed to parenting that involves overstimulation (in the form of exaggerated demands and expectations, punishment and threats, maltreatment, sexual or emotional abuse) or hypostimulation (in the form of parental indifference, negligence or abandonment).

As stated before, the child's essential motivation is not only to create attachment relationships, but also to maintain them in order to survive and find some sort of regulation in the environment (Stolorow 2011). Therefore, children exposed to harmful parenting will tend to adopt strategies that aim to increase caregivers' availability and, at the same time, reduce the probability of aggression from them.

Such strategies may be grouped into two categories (Crittenden 2013):

A. *Coercive* strategies, which involve the magnification of negative emotional states, as a way to regain or maintain the caregiver's availability. Some behaviors included in this category are impulsiveness, risk-taking, tantrums, feigned helplessness, repeated emotional demands, etc. They involve an increase in attention to self-states (*how I feel, what I need*), as well as a relative neglect of external information (*how the other feels, what they need*). These strategies usually emerge in response to inconsistent parenting practices. For example, caregivers who are depressed or overwhelmed by stress, tend to approach and move away from their children in unpredictable ways, something that compromises the child's ability to predict distressing separations and to prevent them. Faced with this difficulty in learning relational contingencies (*what leads to what*), the child systematically magnifies her emotional states, in an attempt to control caregivers and to keep them close.

B. *Compulsive* strategies, which involve the minimization of negative emotional states, as a way to maintain the caregiver's availability and reduce the probability of aggression. Behaviors included in this category are compliance, pseudo-independence, caregiving behavior, inhibition, among others. They involve a decrease in attention to self-states, as well as heightened attention to information about feelings and needs in others. These strategies usually emerge in response to parenting practices that are harmful and predictable at the same time. In cases where caregivers are systematically rejecting, critical, or invasive, the child develops an organized knowledge of relational contingencies. Faced with this "clarity" as to the precipitants of abuse, these reduce probability of aggression by minimizing their affective signals, becoming "invisible" or compulsively altruistic towards their parents.

In the absence of reparative relationships, these strategies tend to stay and evolve throughout development: they become systematic, they generalize across relationships, and they change in form despite maintaining their function, which constitute a relevant damage/trauma to child's wellbeing and wellbecoming. Finally, they may enter into the realm of parenting. Early coercive strategies (e.g. exaggerated emotional demands) may evolve into crystallized forms of aggression and interpersonal control, with respect to others as well as to one's children. Likewise, early compulsive strategies (e.g. anxious obedience or deference) may evolve into overprotection, with respect to others as well as to one's children.

In this way, the development from early trauma to dysfunctional parenting may show a *pendulum effect* (Crittenden 2013), as seen in the following examples:

> Miguel's father has a coercive parenting style: he is easily frustrated, and often resorts to shouting and threatening. In order not to stir negative feelings in his father, Miguel has become self-sufficient and almost 'invisible': he seldom shows his needs, asks for help or shows distress; also, he tends to play by himself and has a very independent demeanor. With the years, Miguel has grown up into an efficient, but emotionally 'blind' parent: he is distant, shows a tendency to neglect his children's emotional needs, and pushes them towards emotional autonomy.
>
> Ana's parents suffer both from unresolved grief. This makes them inconsistent and unresponsive. Sometimes, they are emotionally numb and seldom available to play or stimulate her, while other times they seem to come to life, as if 'reanimated' by causes that are unknown to Ana. In such situations, she unexpectedly finds herself in the arms of enthusiastic, available parents.

(continued)

With time, Ana develops coercive strategies: she is hyperactive, loud and violent. This helps her regain attention and stimulates parental responses (it matters little to Ana that some of these responses are punishments). Now that she is a parent, Ana often uses practices of a reactive nature: she is often loud, threatening and aggressive towards her child.
Julio's critical, controlling parents seem to be content and become less distressing when Julio behaves with compulsive compliance. With the years, he develops a thoughtful, sensitive and somewhat weak character. This translates into compulsive parenting practices: Julio tends to abdicate of his authority towards his daughter, something that leaves her disoriented with regard to rules, and often alone in the face of difficult emotions.

The pendulum effect is sometimes accompanied by parental discourse focused around inversion of practices: 'My parents just ignored me, I don't want my children to feel that loneliness'; 'My parents were harsh and bitter, I will be kinder towards my son'; 'My parents were soft and I never had any guidance in life: I will show authority and consistency with my daughter'.

Empirical evidence has shown this pattern of discontinuity in the intergenerational transmission of parenting styles and relational trauma (Crittenden et al. 1991; Hautamäki et al. 2010; Shah et al. 2010). This phenomenon raises important concerns with respect to the visibility of harm and the visibility of the mechanisms that lead to harm, which may lead to some ethical open question on social justice and public mental health policies with consequences for social epidemiology.

Concerning the visibility of harm, compulsive and coercive strategies may confuse the professional community in ways that make relational trauma invisible, and leave children in lack of adequate protection. Specifically, coercive strategies may show as disturbing behavior (e.g. impulsivity, aggressiveness, opposition, etc.) that often leads to psychiatric diagnoses (such as ADHD or Oppositional Defiant Disorder). Such diagnostic categories, despite facilitating communication between professionals, lose sight of an essential point: the disorder is not a property of the child, but of the relationship. Children in this category are often referred to individual, behavioral and/or pharmacological treatments, which do not address the real causes of their mental health problems and, in the most serious cases, involve institutional revictimization.

Compulsive strategies, on its part, involve overadaptation and an apparent absence of negative emotionality: these children are thoughtful, careful, and socially facile. The adult world, in its tendency to praise and reward such behaviors, provides positive reinforcement of the strategies. This is harmful for the child's development. Despite concealment, negative emotions do not cease to exist, and in the most severe cases, they may take alternative 'routes' of expression, such as physiological dysfunctions (e.g. sleep or respiratory problems), or occasional outbursts of negative emotion, that hijacks the person's psychic functioning and can lead to unpredictable outcomes.

In relation to the visibility of the mechanisms that lead to harm, the repetition of trauma across generations, when discontinuous/heteromorphic, can be subtle and difficult to detect. A physically abused child can become an overprotective parent. A sexually abused girl may turn into a physically and emotionally distant parent. In cases as these, defining parenting practices as the result of the intergenerational transmission of trauma can be difficult. However, the detection of this intergenerational component would be essential for the adequate protection of these children, the prevention of early interpersonal trauma, and the psychosocial assistance to vulnerable families and, as such, to public health and social justice.

Concretely, the psychological issues associated with the discontinuity in the intergenerational transmission of trauma involve a particular problem of justice where some dimensions of the phenomenon suffer from invisibility. In this sense, applying the concept of hermeneutical injustice to some of the concrete situations involved may be a helpful strategy to shed some light and rebalance approaches on social epidemiology and public health.

The case of children who use compulsive strategies may be a good example of what has been stated. These children behave in ways that result directly from relational trauma: their compliance, autonomy and overdapatation are aimed towards preventing aggression and guaranteeing some sort of availability from caregivers. These behaviors speak about children who have been exposed to harm. However, they are seldom seen as symptoms of relational trauma, or signs of psychological suffering. Rather, the adult world (teachers, parents, and an increasingly performance-oriented society) tends to praise them, thus reinforcing compulsive strategies, which come at a great cost for personal wellbeing and mental health.

As a result, from a collective perspective, these myths and prejudices, as an instance of hermeneutical injustice, may unconsciously modulate the way some results from social epidemiology are interpreted and even what makes it into these studies, what type of issues deserve and receive attention and which ones are not properly taken into consideration.

Paradoxes and Ethical Open Questions

In this section we will address the paradoxes and normative open questions generated as a result of the previous considerations related to public health and the intergenerational transmission of trauma.

Two essential normative questions on what ought to be done by parents and the state would be (a) what should be the focus of public health policies when it comes to mental health and childhood, that is, where should the focus be placed in order to, secondly, discern (b) what type of responsibilities should be assumed by each agent of justice involved (mainly the state and the caregivers).

By approaching these normative uncertainties applying the idea of hermeneutical injustice we hope to enrich the debate and contribute with some suggestions.

A Question on the Focus: Public Mental Health Policies in the Face of Trauma

With regard to the focus, that is, what should be prioritized; at least two perspectives could be found, depending on whether the aim is to eradicate pathologies and problematic behaviors, or to improve children's wellbeing. Although the results might be similar in some cases, the election of one of these perspectives could perpetuate the invisibility in the case of heteromorphic transmission of trauma (mostly, when victims develop compulsive attachment strategies).

First, if the focus is centered on the prevalence or the extinction of a concrete type of pathology or behavioral problem, rather than on the person's wellbeing as a whole, the lack of awareness of hermeneutical injustices related to children's mental health may lead to a paradoxical situation from the point of view of social epidemiology. It might be the case that the decrease in the figures related to the occurrence of a public health's problem becomes an indicator of a fact that enables the occurrence of other issues, equally relevant for the next generation's (psychological) wellbeing. Moreover, this could be ethically aggravated by the lack of social recognition as an instance of injustice (hermeneutical injustice), and could involve the imposition and perpetuation of an epistemic limitation consisting in the inability to establish the link between the different intergenerational contexts and the transformation of an instance of pathology into another one.

As previously mentioned, an example beyond the chosen issue could also be found in the cases of sexually abused victims who become negligent parents by overprotecting or by being emotionally unavailable parents, so that the second generation's psychological problems do not make it into the official data on the prevalence of sexual abuse and may not be considered as a case of maltreatment as it is socially perceived as not so traumatic compared to the first generation's one.

Another example could be found in the convergence between drug-addict caregivers and excessively responsible children, who develop psychosomatic conditions. By way of example, sleep problems and asthma—the latter often considered a psychosomatic disease—appear to be influenced by the socio-economic status of the child (Williams et al. 2009).

Likewise, if the focus is centered on the child behavior or the pathology, ignoring the intergenerational perspective, the real mechanisms of the child's behavior could go unnoticed or could be inadequately treated, something that would entail a double victimization.

On the contrary, viewing relational trauma and its transmission within a framework of discontinuity, public health policies could better understand how the effects of trauma may unfold in unexpected ways. Trauma effects may be *transformed*, as has already been shown in detail. Trauma effects may also be *dormant* for months and years (for example, compulsive strategies during the school years may seem highly adaptive, whereas by adolescence and the beginnings of close affective-sexual relationships, they may be an important source of suffering and maladjustment). In this sense, we could argue that the effects of trauma are *intermittent* in their manifestation (or, at least, in their conspicuousness), thus obscuring further the visibility of harm and of its mechanisms.

In this sense, the lack of a child-centered perspective involves the risk of falling into certain myopia, leaving a segment of population without the required tools to conceptualize and recognize a problem of such developmental and life impact. Thus, although public health policies may indeed reduce the prevalence of child physical maltreatment and contribute to the reduction of social inequalities, paradoxically, if the analysis is not further developed and the intergenerational relationships are not well examined, hermeneutical injustices may be perpetuated, (involving the lack of protection, recognition and proper study of the next generation), even despite the decrease in the figures concerning the prevalence of such instances of abuse. Therefore, a perspective centered on the child/person, rather than on the behavior, alongside the discontinuity framework, could have a positive impact on both, public mental health measures and the recognition and eradication of hermeneutical injustices.

One last issue concerning the focus, deeply connected to the dissonance between manifest and operative concepts and hermeneutical injustice, is the usage of the terms maltreatment or trauma, as it was already introduced in section "Justice, Public Health, and the Moral Expectation of Recognition".

From the child's perspective in every generation, it is easy to see how there is a leap from traumatized children to traumatized children, even though the first generation were traumatized by a violent physical abuse and the second one by a subtle but chronic attack to their personal identity, to mention an example.

However, if the focus is on the adult's behavior and parental responsibility, as we will develop in the next section, then the language will shift from "trauma" to "maltreatment", "aggressor" or "abusive caregiver". In other words, if the subtle cases are accepted as a problem of public health and justice, and if those children are considered as being exposed to relational danger and trauma—to a greater or lesser extend-, then that given society could refer to those caregivers as harmful and to those cases as cases of maltreatment. This creates a dissonance between manifest and operative concepts on child maltreatment, which, in turn, generates the mentioned invisibility in the cases of hermeneutical injustices.

To put it differently, if the focus is placed on the adult, that society would jump from physical maltreatment to psychological maltreatment from one generation to another, which leads to one of the main reasons for hermeneutical injustice and social disagreements. Naming those cases as an instance of intrafamily maltreatment—even if the child suffers from a subtle instance of it-, is socially perceived as an overestimation. Likewise, categorizing those caregivers as abusers is also perceived as inflated, especially from the point of view of the adult community, which may perceive those symptoms as irrelevant.

In the same way as Jenkins (2016) denounces the prevalence of hermeneutical injustices and a dissonance between manifest and operative concepts in the contexts where spousal rape is not considered as such; it is still perceived as nearly irrelevant and far from being a case of abusive parenting if the child is ignored, verbally ridiculed o unvoiced in any initiative. This would show how the operative concept of maltreatment refers mainly to the most severe cases of physical and sexual maltreatment, letting so the cases of psychological maltreatment obscured.

Hence, the ways hermeneutical injustices are involved in subtle traumas during childhood are directly connected to how parenting styles and different traumas are perceived. In fact, "there is a weaker societal consensus about how to distinguish emotional abuse from suboptimal parenting" (Trickett et al. 2009, p. 28) or whether suboptimal parenting is actually a type of emotional abuse.

As a result, the social acceptance of the fact that there is an injustice and a problem of public health in the subtle cases would turn some unskilled caregivers into the controverted category of abusive or at least harmful caregivers. This, as we will address in the following section, leads to another set of normative questions in terms of children mental health and parental and state responsibility.

A Question on Responsibilities: Victims Taking Care of Victims

In terms of responsibility, dilemmas would arise regarding two issues. First, the double role of parents as children from the previous generation, and as agents of justices with parental responsibilities toward the new generation. Second, the difficulties in measuring what policies should be prioritized by the state as the second main agent of justice for children.

In relation to the first question, the main caregivers would be at the same time the victims of a previous situation and, as a new agent of justice toward that child, responsible of the intergenerational transmission of trauma, either repeating the same type of abuse, or producing discontinuous forms of harm upon the child's development. Hence, once that child has become an adult with parental responsibilities, a dilemma on what ought to be demanded from him/her arises.

To put it differently, what can be demanded to a victim of some kind of abuse or maltreatment clashes with what can—and should—be demanded to that person as an agent of justice. Therefore, it is important to tackle some normative uncertainties generated by the discontinuity of intergenerational trauma transmission on its own. Should immediate actions be implemented such as obliging the adult to cure the problems associated with attachment as a parental duty? Can society provide the means to protect children who are exposed to relational danger and, at the same time, provide care to traumatized parents? Could these two aspects be considered as part of a unitary process aiming at the eradication of trauma and its transmission?

In relation now to the second question, should then public health policies focus on providing the caregiver with the necessary tools to ensure a healthy bond, should they only focus on the elimination of the most unhealthy parenting styles without necessarily providing the extra resources needed to generate a good one, or should they focus on providing the child with the tools to minimize, alleviate and overcome such a situation? How should such policies be designed so that they do not stigmatize the caregivers without sacrificing an important aspect of child's mental health and development?

In order to answer these questions, and now from the point of view of hermeneutical injustice, it is crucial to highlight that, especially in these issues of public mental health, "the victim is unaware of the fact that she is instrumental in reproducing and propagating the same injustice" (Bufacchi 2016, p. 9), even through different types of abuse. The child, now as an adult, may turn into a perpetrator of that injustice to someone more vulnerable (the next generation) by failing to prevent, recognize, or by being indifferent to the damage inflicted, so that the inculcated injustice would run invisible across generations until a critical evaluation of the event is socially assimilated, as it happened concerning other hermeneutical injustices on race and gender.

Moreover, given the fact that in this case the perpetrator is also a victim and that the attachment between the child and the parents is crucial to the child's wellbeing and development, it is also remarkable how the traumas addressed in this chapter are especially paradoxical: contrary to other cases, restraining measures between the victim and the perpetrator would involve more negative effects to the victim. This, in turn, makes especially challenging to concrete the scope of the parental and state responsibility and to design public health approaches that eradicated the inculcated hermeneutical injustices while making justice to all the subjects involved without automatically criminalizing all forms of suboptimal parenting.

These are complex questions that warrant reflection and discussion. The simultaneous condition of victim and perpetrator that characterizes many dysfunctional parents demands a complex response from clinicians, policy-makers and the society as a whole. However, we believe that measures will only be successful if they take place within a framework of *ecological transmission of security*. This idea involves that prevention and intervention in the face of early relational trauma should always include the care of parents, despite the fact that they may be harmful to the child. This becomes clearer in light of epistemic injustice: a perpetuated instance of hermeneutical injustice will not be visible and eradicated until perpetrators and spectators become aware of the existence of such. Parents will not function as safe caregivers until they have earned an experience of security, something that the agents of justice (the state, the institutions and healthcare professionals) are responsible of providing. Therefore, trauma-related interventions should move away from educational and corrective approaches (e.g. parental education, parental training, behavior modification) and move towards a more reparative approach given the depth and the invisibility of the injustice. This would involve, among others: the establishment and monitoring of a trust-based therapeutic alliance with parents (vs. a judgmental stance); the management and regulation of parents' difficult emotions; the development of strength-based interventions (vs. interventions exclusively focused on deficits and dysfunction); or the validation of parents as legitimate caregivers.

For clinicians and professionals, this would mean to function as *transitional attachment figures* (Crittenden 2013) for parents. For policy-makers, this would mean to invest in systems of care that are not reactive but preventive, not coercive but empowering of parents, and complex in the allocation of responsibility. For the society as a whole, this would involve the incremental construction of a culture of care and informed compassion that would destroy the "active ignorance" required to perpetuate an instance of hermeneutical injustice. Finally, for the state as an agent of justice, this would imply the implementation of an intergenerational perspective when it comes to public mental health. As a result of all this, experiences of security earned by parents through intervention may 'spill' into the relationship with the child.

Preventive measures within the discontinuity framework would have an impact on society as a whole and on its collective imaginary on childhood and parenting. Bringing back the discourse of hermeneutical injustice, a preventive approach on public mental health and intergenerational trauma transmission would deactivate what has been named as "active ignorance", meaning all the epistemic attitudes and vices that contribute to sustain and expand hermeneutical injustices, such as arrogance, laziness and close mindedness, especially among the subjects privileged by the unjust situation (Medina 2013, pp. 30–39). In this case, this would affect the spectator (society) and the victims when they become perpetrators.

"Structural active ignorance can be corrected only by developing epistemic virtues such as epistemic humility, curiosity/diligence, and open-mindedness" (Medina 2013, p. 23), which are virtues that preventive measures would develop in caregivers. In fact, hermeneutical injustices constitute a case of lack of moral curiosity and sensitivity towards someone else's experiences and views, and affect both the victims (who may not even know they are victims), and their social environment (which may misunderstand or trivialize the injustice). To overcome it, interactions with significantly different epistemic others is required, as well as recognition and engagement, which is not possible without implementing the previously suggested preventive measures.

Finally, the problem of responsibilities and priorities irremediably leads to the classic questions on thresholds, especially if resources are limited: should policies concentrate on the most severe cases, although the prevalence of these ones may be not so high, or, on the contrary, should they focus on the most extended public health issues affecting a large number of citizens, although the person may be able to develop relatively well compared to the most severe cases.

Here a sufficiency rule of distribution is often understood as the best solution in terms of social justice (Nussbaum 2011). This would imply that children's development should be guaranteed at least to reach a sufficient level of wellbeing. Nevertheless, the fact that addressing subtle cases is sometimes perceived as beyond the wellbeing's thresholds needed to ensure a minimum level of social justice and health may actually be an effect of the hermeneutical injustice and the active ignorance involved in the social perception of these cases, especially when it comes to evaluate the damage and long-terms effects on a person's development.

Moreover, the fact that the measures needed, if these problems were taking into consideration, would be mainly preventive, and therefore cost-effective, would lead to question the premise of the scarcity of resources and how the collective imaginary (active ignorance) and the hermeneutical injustice is modulating the perception of both, the case and the responsibilities in its solution.

Discrediting someone's experience of harm by comparing it with other socially recognized cases, setting so the focus on the more invisible effects instead of on the person, comparing different types of harm to a point where the result of the comparison would leave some experiences obscured are in fact mechanisms of epistemic injustices that surely shape how public health issues and institutional responsibilities are formulated.

In other words, it is impossible to give an answer on what should be done in terms of justice without concrete knowledge on the case and an evaluation of the social context and circumstances. What a person may need in terms of healthcare to reach a sufficient level of wellbeing may differ from case to case. Although both claims of justice, what a person suffering from a coercive parenting style would need would differ from what a person victim of physical maltreatment would required in order to reach that sufficient threshold of wellbeing. Thus, the question of choosing whose basic rights should be brought to reality would turn to be perverse, in the same way in which comparing the needs of a newborn and an adolescence in order to prioritize just one group would also be deeply unjust. In a nutshell, the fact that some comparisons are perceived as absurd and shock us while some others are not even detected is actually a sign of the presence of an epistemic injustice. Thus, it will not be possible for children to reach a fair wellbeing threshold if the primary causes for their conditions are not even perceived as relevant and changeable because they are not even socially recognized as such.

That being said, once again, prevention may be a suitable response for both, the public health problem and the socially inculcated bias related to it. Instead of reacting to mental health problems that are already crystallized and/or to overt parental abuse, resources may be allocated to preventive services within the primary mental health system. This would entail the sensitization of professionals (nurses, midwives, social workers, doctors, etc.) who work in maternity services, pediatrics, daycare, or community health centers, among others. With relatively little expenditure, these professionals may be trained to detect emerging relational difficulties between caregivers and children, to make referrals when needed, and follow up. Attachment is universal: in order to support it, clinical approaches are not needed (clinical approaches are suitable for the most severe cases, those where attachment does not need to be supported, but rather repaired or interrupted, if the child's survival is at stake). In most instances, vulnerable attachment relationships may be protected with short-term, low-intensity interventions (instead of protracted clinical treatments) as long as these are administered *in time*.

Thus, instead of swinging from approaches that leave the subtle cases under invisibility to those perspectives that understand parental responsibility in a way that would lead to a massive criminalization of suboptimal parental styles, we suggest that understanding the problem as a case of hermeneutical injustice and from the framework of discontinuity would help to see preventive ecological measures not just as desirable or supererogatory claims, but as a basic requirement in terms of justice.

Conclusions

We have presented how taking into account the social perpetuation and transmissions of several aspect of some mental health problems during childhood and the inherent injustice in them may lead to reformulate the reactive and preventing measures in a more human way without renouncing to parental and state responsibility toward children. Likewise, in order to overcome current instances of injustice related to health and social inequality, interdisciplinary collaboration between scientific and philosophical fields seems essential and a fruitful path in order to successfully untangle epistemic challenges and practical dilemmas.

In this sense, and supported by the theory of epistemic injustice, we have advocated for health and intervention policies that focus on prevention of relational trauma within the family, rather than on the eradication of negative behavior in children. This attachment-centered, preventive approach would entail a developmental framework that is sensitive to the flexible patterning of parent-child relationships across generations, and to the fact that, underlying the diversity of relational strategies and parenting practices, there is a basic, universal need to protect oneself against isolation and harm.

An approach to the assessment of mental health problems and psychosocial well-being in children that is not focused on behavior or symptoms, but rather on *function* would lead to the core of some attitudes that would perpetuate the hermeneutical injustices such as an overestimation of knowledge or lack of curiosity on the victims. By questioning what strategy underlies the child's behavior or what the relational context within which the symptoms have developed is, coercive and compulsive strategies underlying manifest problems would become visible and easier to detect, which would increase the chances to reduce social inequality through public health measures, and therefore the prevalence of hermeneutical injustices in the face of childhood.

This could lead to measures that have been perceived as far from what is expected form public health scope but that, as previously argued, would be key to break with the intergenerational and social perpetuation of trauma and injustice. Something as specific as paying special attention to children that are docile, compliant, or show overadaptation, in order to ascertain if these behaviors are part of a general compulsive strategy to relational trauma, would be an example of the subjected approach, with a considerable deep impact on a given society levels of health and justice in the long-term perspective.

In line with this, we have defended an approach to the treatment of mental health problems in children not centered around behavior or symptom control, but around reducing danger within the child's caregiving environment (i.e. family or school). Likewise, we have defended an approach on parental responsibility beyond discourses of blame and criminalization. Once danger is reduced, attachment can be straightened out spontaneously (i.e. coercive/compulsive strategies may be reorganized into more balanced attachment strategies) (Crittenden 2013).

Beyond the suggested issue on childhood and the transmission of trauma, we are sure that an inclusive preventive and intergenerational methodology as well as a person-centered approach could also be successful in detecting the nature of other instances of epistemic injustice and eradicating other forms of social injustice related to public health and inequality that suffer from invisibility. Once applied to public health, the acknowledgment of the injustices and asymmetries in power and visibility may lead to new ethical commitments in terms of responsibility, beyond causal explanations and culpability, and slightly different methodological paths to approach recalcitrant social and health problems.

We hope that the critical analysis of social and mental health problems combining psychological tools (attachment theory, discontinuity framework) and philosophical ones (epistemic injustice, structural active ignorance) could contribution to find practical directions in order to improve a society levels of justice, health and well-being.

References

Améry, J. (2001). *Más allá de la culpa y de la expiación*. Valencia: Pre-textos.

Berlin, L. J., Appleyard, K., & Dodge, K. A. (2011). Intergenerational continuity in child maltreatment: Mediating mechanisms and implications for prevention. *Child Development, 82*(1), 162–176.

Bowlby, J. (1969). *Attachment and loss: Vol. 1: Loss*. New York: Basic Books.

Bufacchi, V. (2016). Love in the time of epistemic injustice. In M. Marinucci (Ed.), *Jane Austen and philosophy* (pp. 3–13). New York: Rowman & Littlefield.

Corbí, J. (2005). Emociones morales en la flecha del tiempo: Un esquema de la experiencia del daño. *Azafea. Revista de filosofía, 7*, 47–64.

Crittenden, P. M. (2013). *Raising parents: Attachment, parenting and child safety*. London: Routledge.

Crittenden, P. M., Partridge, M. F., & Claussen, A. H. (1991). Family patterns of relationship in normative and dysfunctional families. *Development and Psychopathology, 3*(4), 491–512.

Fricker, M. (2007). *Epistemic injustice: Power and the ethics of knowing*. New York: Oxford University Press.

Haslanger, S. (2012). What are we talking about?: The semantics and politics of social kinds. In S. Haslanger (Ed.), *Resisting reality: Social construction and social critique* (pp. 365–380). Oxford: Oxford University Press.

Hautamäki, A., Hautamäki, L., Neuvonen, L., & Maliniemi-Piispanen, S. (2010). Transmission of attachment across three generations: Continuity and reversal. *Clinical Child Psychology and Psychiatry, 15*(3), 347–354.

Jenkins, K. (2016). Rape myths and domestic abuse myths as hermeneutical injustices. *Journal of Applied Philosophy, 34*(2), 191–205.

Medina, J. (2013). *The epistemology of resistance: Gender and racial oppression, epistemic injustice, and resistant imagination*. Oxford: Oxford University Press.

Mullin, A. (2014). Children, vulnerability, and emotional harm. In C. Mackenzie, W. Rogers, & S. Dodds (Eds.), *Vulnerability* (pp. 266–287). New York: Oxford University Press.

Nussbaum, M. C. (2011). *Creating capabilities: The human development approach*. Cambridge, MA: Belknap Press of Harvard University Press.

Pears, K. C., & Capaldi, D. M. (2001). Intergenerational transmission of abuse: A two-generational prospective study of an at-risk sample. *Child Abuse & Neglect, 25*(11), 1439–1461.

Powell, B., Cooper, G., Hoffman, K., & Marvin, B. (2013). *The circle of security intervention: Enhancing attachment in early parent-child relationships*. New York: Guilford Publications.

Schore, A. N. (2010). Relational trauma and the developing right brain: The neurobiology of broken attachment bonds. In T. Baradon (Ed.), *Relational trauma in infancy: Psychoanalytic, attachment and neuropsychological contributions to parent–infant psychotherapy* (pp. 19–47). New York: Routledge/Taylor & Francis.

Shah, P. E., Fonagy, P., & Strathearn, L. (2010). Is attachment transmitted across generations? The plot thickens. *Clinical Child Psychology and Psychiatry, 15*(3), 329–345.

Stolorow, R. D. (2011). *World, affectivity, trauma: Heidegger and post-Cartesian psychoanalysis*. New York: Routledge.

Trickett, P. K., Mennen, F. E., Kim, K., & Sang, J. (2009). Emotional abuse in a sample of multiply maltreated, urban young adolescents: Issues of definition and identification. *Child Abuse & Neglect, 33*(1), 27–35.

Widom, C. S., Czaja, S. J., & DuMont, K. A. (2015). Intergenerational transmission of child abuse and neglect: Real or detection bias? *Science, 347*(6229), 1480–1485.

Williams, D. R., Sternthal, M., & Wright, R. J. (2009). Social determinants: Taking the social context of asthma seriously. *Pediatrics, 123*, S174–S184.

Social Determinants of Health: Citizenship Rights and Responsibilities

Oriol Farrés-Juste

Abstract Social determinants of health (SDH) include all kinds of social, political and economic factors that have an influence on people's health, and they are the main drivers of inequities, that is, unjust inequalities in health. According to studies of SDH, there is a clear social gradient in health, which raises a number of ethical and political questions that we must consider carefully. From the standpoint of citizenship, the redistribution of SDH is a matter of both rights and responsibilities. Scientific (descriptive) analysis of SDH must be put in relation to value-oriented (normative) research on bioethics, theories of justice, and human rights. Finally, SDH and population health are directly related to the rights and duties of citizens. Solidarity (with others) and responsibility (for oneself) are two sides of the same coin in order to improve public health.

Keywords Social determinants of health · Health inequities · Citizenship · Justice · Human rights

Introduction

According to Marmot, "there had been a curious disjunction between the fact of the social gradient (in health) and the belief that the major killer in developed countries, coronary heart disease, affected most those of high status. The view appeared to be that there were diseases of the rich (heart disease and cancer) and diseases of the poor. Mortality of all causes reflected a balance of the two. (. . .) We know that this is inaccurate." (Marmot and Wilkinson 2006: 2) The emphasis on this disjunction between facts and beliefs is very useful in order to take into consideration the relationship between ideology and social science. What social epidemiologists Marmot and Wilkinson's 'solid facts' show is that the poor have more health problems (diseases and shorter life expectancy) than the rich. The crucial matter

O. Farrés-Juste (✉)
Department of Philosophy, Universitat Autònoma de Barcelona (UAB), Bellaterra, Spain
e-mail: oriol.farres@uab.cat

J. Vallverdú et al. (eds.), *Philosophical and Methodological Debates in Public Health*, https://doi.org/10.1007/978-3-030-28626-2_9

here is to think about the normative dimension of this problem. So there are two problems: (1) there is a discussion of the facts: whether or not SDH studies are accurate enough and the extent to which they are valid; and (2) given the fact that there are inequalities in health depending on social status (expressed in SDH), there will also be a discussion about values, that is, a debate about the normative question: whether this situation is fair or not, and how it can be changed if it is indeed really unjust.

This chapter deals with the second subject matter: the normative dimension of SDH. Since "Epidemiology and Social Justice in the Light of Social Determinants of Health Research" (Venkatapuram and Marmot 2009), it is clear that research into SDH must be inter-disciplinary and must make room for Moral and Political Philosophy. What's more, the theories of justice in the Rawlsian tradition are especially useful for considering the lack of fairness in health inequalities, which are based on the social positions and economic situations of the individuals involved. If the poor are in worse health than the rich, then they are in a position to make justified demands regarding their needs and interests. In a word, as Puyol (2012: 178) establishes, studies of SDH have changed the conception of equity, public health and the consideration of the principles of bioethics, such as beneficence, non maleficence, autonomy and justice (contrary to the bioethical tradition, justice would be even more prominent than autonomy in some cases).

As Marmot and Venkatapuram state, "epidemiologists and philosophers need to create a meaningful exchange of ideas to ensure that the reasoning used in epidemiology is defensible, both for the instrumental moral purpose of epidemiology to be clarified, and for conceptions of social justice to be more relevant and coherent." (Venkatapuram and Marmot 2009: 89) This is very important in order to show that there is a way for, according to the scientists that carry out studies of SDH, philosophy and science to be able to cooperate in order to reflect upon the social conditions and the health inequalities that these social conditions produce on a daily basis.

This necessary reflection has two different sides: on the one hand, there is a field of study of causal inferences in the analysis of SDH, so there is a need to reflect on causality applied to health issues in terms of a philosophy of medicine (or strictly speaking, "a philosophy of epidemiology"), as Alex Broadbent (2013) has recently shown in his studies of inductive inferences in epidemiology. According to Venkatapuram and Marmot, "because the ontology of social determinants is still being identified, social epidemiologists need to be especially diligent about making causal inferences between social distribution patterns and social features. Indeed, this need for social epidemiologists to give more attention to the logic of the reasoning used to infer causation provides an opportunity for epidemiology as a whole to undertake a thoroughgoing analysis of the reasoning used in making causal inferences." (Venkatapuram and Marmot 2009: 85) But on the other hand, the normative issues of justice, redistribution, and inequities applied to health constitute an emergent field of study for the future of bioethics. This chapter aims to describe it.

Social Epidemiology and Social Justice

The discovery of the relationship between health problems and social conditions is not recent (Aloisi et al. 1968). Leaving aside Hippocrates, Bernardino Ramazzini published "De Morbis Artificium Diatriba" in 1713, a pioneering study in which he sought to define the types of diseases characteristic of different labour conditions. Nineteenth century researchers Virchow, Neumann and Lebuscher reconfirmed the influence of social and environmental factors on the aetiology of disease. They believed in the importance of preventive measures to control them. Chadwick and Shattuck also emphasized the importance of social factors in disease prevention and control programs. There were various revisions: Antonovsky evidenced the relationship between disease and economic situation; Kitagawa and Hauster observed that mortality rates are significantly higher among the lower classes; and Stevenson proposed the analysis of death rates by classifying social classes into six groups. But the publication of the Black Report (Gray 1982) in 1980 was the starting point of new social epidemiology.

Although in a sense it is certainly true that all epidemiology is 'social' epidemiology, the scope of so-called 'social epidemiology' is the health inequalities among populations. This cannot be said about all the other kinds of epidemiology (Berkman and Kawachi 2000). Research of various determinants of health can lead to results where we find that natural, geographic or even cultural factors are determinant in certain diseases. However, it is likewise true that the information provided by social determinants of health is impressive in terms of the "causes of the causes" (Marmot and Wilkinson 2006) of diseases, that is, in terms of the social distribution of disease. The expression 'causes of the causes' means that the direct causes of disease (indicated by way of biological markers) can have a social determinant that has an influence on the occurrence of such a direct cause. "We go further to examine the causes of these causes. It is not an accident that people consume diets high in saturated fat and salt. It represents the nature of the food supply, culture, affordability, and availability among other influences. These are the causes of the causes. For example, given that smoking is such an important cause of premature disease and death, we need to understand the social determinants of smoking. In particular, in many rich countries now, there is a social gradient in smoking: the lower the socio-economic position, the higher the rate of smoking." (Marmot and Wilkinson 2006: 3)

This is not a new idea. Marx and Engels, in the origins of social sciences, had something similar in mind. Marx explains: "Dr. Edward Smith, during the cotton crisis caused by the American Civil War, was sent by the English Government to Lancashire, Cheshire, and other places, to report on the sanitary condition of the cotton operatives. He reported that from a hygienic point of view, and apart from the banishment of the operatives from the factory atmosphere, the crisis had several advantages. The women now had sufficient leisure to give their infants the breast, instead of poisoning them with 'Godfrey's cordial'." But what was Godfrey's cordial? We can find the answer to this question in Engel's words: "One of the most injurious of these patent medicines is a drink prepared with opiates, chiefly laudanum, under the name Godfrey's Cordial. Women who work at home, and have their own and other people's children to take care of, give them this drink to keep them quiet, and, as many believe, to strengthen them. They often begin to give this medicine to newly born children, and continue, without knowing the effects of this 'heart's-ease', until the children die. The less susceptible the child's system to the action of the opium, the greater the quantities administered. When the cordial ceases to act, laudanum alone is given, often to the extent of fifteen to twenty drops at a dose. The Coroner of Nottingham testified before a Parliamentary Commission that one apothecary had, according to his own statement, used thirteen hundredweight of treacle in 1 year in the preparation of Godfrey's Cordial. The effects upon the children so treated may be readily imagined. They are pale, feeble, wilted, and usually die before completing the second year. The use of this cordial is very extensive in all great towns and industrial districts in the kingdom."

In the present-day terminology of SDH, the 'direct cause' of these diseases was laudanum, of course (and it could be indicated by means of biological markers). However, the 'cause of the cause' of these terrible diseases and deaths among children (because of the poison called 'Godfrey's Cordial') was the social, economic and labour conditions of their mothers at that time. Acting directly on the social conditions might have meant a change in women's labour conditions. Then, this change will have produced a decrease in this health problem among children. So it is not really easy to cleanly separate public health from social justice. They are interrelated. As Marx explains, these incidences ended with the transformation of labour conditions due (paradoxically) to a crisis. One "pathway" (Marmot and Wilkinson 2006: 3) by which the social environment affects health is through the influence of material conditions (the actual circumstances in which people live and work). Studies of SDH are very clear in this respect.

Therefore, in pragmatic terms, it would be useful to detect and intervene on these SDH: that is, to directly attack the 'causes of the causes'. Praxiology teaches us that we need to organise certain means in order to achieve specific goals on technical grounds (Kotarbinski 1965). From this perspective, social epidemiology is an applied science committed to a social intervention in order to obtain programmed results, whereby "an approach to the new epidemiology stresses that epidemiologists need to become activists in applying their findings to develop health policy and manage health care resources. More and more, the new epidemiology has become social epidemiology." (Cwikel 2006: 90) Epidemiology leads to social epidemiology; and then, social epidemiology should lead to praxiologically efficient social intervention.

What kind of intervention is needed in order to conceive activism in social epidemiology? The answer is redistribution (Upshur 2002). The transformation of social determinants of health means redistributing economic resources and opportunities, material goods and assets. The social epidemiologist, according to Cwikel's paradigm, becomes an activist who raises concerns in favour of particular public policies. As an advisor, he is in a way somewhat close to an 'activist'. This is so because once we have understood that social inequalities—and not just poverty—are the causes of bad health in vulnerable populations, a measure of income redistribution becomes a measure of public health. This is why we will need a normative criterion to justify the implementation of such political measures. Rawlsian theory of justice, in a sense, furnishes this criterion. And furthermore, Habermasian ethics of discourse can also furnish this kind of moral justification. Nowadays, these two approaches are the mainstream of moral theory. Therefore, the combination of both ethical approaches—Rawlsian and Habermasian—can bring about a basis upon which we can raise valid demands for redistribution in terms of public health.

First, the key concept of this ethical turn in social epidemiology is the idea of citizenship. In both Habermasian and Rawlsian theories the conception of citizenship has republican roots, that is, citizens are conceived to be free and equal. At the same time, this republican conception of citizenship is not only formal. There are also material requirements that have to be met. This is the case, for example, of Rawl's theory of "primary goods" (Rawls 1999), according to which every citizen of a well-ordered society must be in possession of certain primary goods that are ultimately the grounds for his or her self-respect as a citizen. Among these primary goods, health and education play a very important role, as do income and wealth. Health plays a vital role in this kind of issue because it is often a condition for the possibility of many other goods and assets. For example, it is not possible to enjoy civil or political rights (or even to have a job) without certain guarantees of good health—or without certain guarantees of a system that provides healthcare and a way to deal with ill-health. In Habermasian terms (Habermas 1991), those who are more vulnerable to ill-health should have an equal opportunity to present their valid demands in the public sphere. Norman Daniels (2008) and Amartya Sen (2002) have shown and developed this dimension of the problem of health policies.

Second, it is important to note that "without a readily available general theory of social justice that is coherent and commonly accepted, we are presently left with relying on a casuistic approach to ethical analysis of health concerns and their relation to the just functioning of other social spheres." (Venkatapuram and Marmot 2009: 86) Although we do not have a general theory of social justice yet, action is imperative. This means that the elements of the Rawlsian and Habermasian theories of justice at hand are to be used in particular circumstances and concrete problems. The 'casuistic approach' is also necessary in the field of health inequities. In fact, the whole tradition of Bioethics as a discipline, since the work by Toulmin and Jonsen (2009), on the one hand, and Jonsen et al. (2001), on the other, has its roots in a casuistic way of reasoning. So the time has come to put forward and apply the casuistic approach of bioethics to study and action with regard to social determinants of health (with a focus on social environment).

Third and finally, the theories of social justice applied to SDH cannot ignore the World Health Organization's (WHO) theoretical proposition about "health inequities" (Whitehead 1990: 29). First of all, it is useful to remember that inequality and inequity are not quite the same. There could be some inequalities that are not equities. The normative criterion in this case is the term "inequity", which means unfair or unjust inequalities (Daniels et al. 2007: 212). According to the WHO, health inequity will depend on three conditions: impairments and mortality that are

(1) avoidable (2) unnecessary and (3) unfair and unjust. These three conditions work in the following manner: those inequalities that are at the same time avoidable, unnecessary and unjust will be considered 'inequities'. "In any case, once a set of impairments/mortality cases are determined to be avoidable, the next step entails identifying which of those cases are unnecessary. And after that, the last criterion determines which of the remaining impairments/mortality are unfair/unjust." (Venkatapuram and Marmot 2009: 87)

The junction of social epidemiology and the theories of social justice may imply a problem of biases in this area of research. Many epidemiologists have dealt with this objection. One of the responses argues that acknowledgement of values in this kind of research does not necessarily entail a lack of objectivity. "An undeniable benefit of a value-free approach to science is the ability to claim that the research is free of potential biases that might come with being explicit about values. However, acknowledging the value base of research does not make objectivity unobtainable." (O'Campo and Dunn 2012: 60) Moreover, given the fact that a value-free social science can itself be a bias, in the end it becomes a matter of trying to achieve a 'reflexive science' or 'reflexivity in science'. This principle has the following characteristics in practice: "A reflexive social epidemiologist researcher would not only analyse statistical data on differential health outcomes but also consider the categories of analysis used and the conditions of data gathering. The combination of these reflexive stances, one focused on the individual researcher and the other on a discipline, could help to explicate underlying value structures." (O'Campo and Dunn 2012: 61) Among others, the Frankfurt School in Philosophy and Pierre Bourdieu in Sociology are outstanding examples of what is meant by 'reflexivity' in the practice of science. That is, this kind of 'reflexivity' is an asset of Philosophy and it is something through which the gap between sciences and humanities can be bridged. We must not forget that bioethics, as a "bridge to the future" (Potter 1971), has always had the goal of combining sciences and humanities. This is important in order to achieve socially situated Moral and Political Philosophy and, at the same time, value-oriented epidemiology.

Public Health and Citizenship Rights

There can be no doubt that social organization has a significant effect on population health. Since its origins, public health has referred to "the science and art of preventing disease, prolonging life and promoting human health through organized efforts and informed choices of society, organizations, public and private, communities and individuals." (Winslow 1920: 191) Here, the key idea is prevention. According to the findings of SDH research, "we can extend the conceptual framework to provide a public health view in which the emphasis is on prevention rather than cure." (Marmot and Wilkinson 2006: 8)

The social hygiene movement is a very good example of positive results in the emphasis on prevention rather than cure. Those major advancements constituted a condition for the possibility of major improvements in health. Universal access to a health system is also important. However, the improvement of living conditions is the key element in terms of public health (Puyol 2012: 179), and means a just redistribution of social determinants of health. The theories of social justice are especially useful in order to justify these necessary political interventions. According to Rawls, for instance, every citizen needs a series of primary goods in order to be a full member of a well-ordered society (income and wealth are typical examples of such primary goods). Health is considered in *A Theory of Justice* to be a "natural primary good". The fact is that health impairments can produce inequalities in access to various social positions, which should be formally open to all members of a just society. Thus, an inequality becomes an inequity. That is why a well-ordered society needs to plan a health system or a way to solve this source of inequities.

It is well worth considering how health inequities produce inequities in other social fields, such as political rights or economic opportunities. But on the other hand, social inequities (expressed in SDH) do indeed produce health inequities. The mutual dependence of these factors deserves some attention. The redistribution of SDH is a way of acting at the same time on the health level and on the justice level of society as a whole. This just redistribution of SDH is a way of preventing diseases and is obviously a way of increasing population health too. So the redistribution of SDH is a public health policy that must accompany universal access to a health system for all members of society (Puyol 2012: 180). However, as shown earlier, preventing is more important than curing; so the emphasis should be put on prevention rather than on cure. Equality of opportunity is the value at stake in this case: the just redistribution of SDH reinforces the principle of equal opportunity for all.

In the political tradition of republicanism (Pettit 1997), citizenship means having rights to take part in public life and to avoid domination. Ill-health is a source of vulnerability. In this sense, the right to health becomes a fundamental piece of the core characteristics of an active citizenship. Regardless of origins, ideology, language, culture, colour or creed, the right to health is a universal feature of social rights, the second wave of rights according to Marshall and Bottomore (1992). No citizenship in the true sense of the word is even conceivable without a reference to the right to health. In particular, it implies "the right to the enjoyment of the highest attainable standard of physical and mental health" (in the words of the constitution of the World Health Organization back in 1946). There is a social responsibility and a series of obligations to satisfy this demand. Moreover, the right to health is recognized as a human right. But how can the right to health be secured?

The answer to this question has to do with States; and in effect, it has to do with social policies of States. In the current world order, States have the ultimate responsibility of implementing the necessary measures to secure the right to health in their territories. "The right to health is relevant to all States: every State has ratified at least one international human rights treaty recognizing the right to health. Moreover, States have committed themselves to protecting this right through international declarations, domestic legislation and policies, and at international conferences." (World Health Organization 2008: 1) Needless to say, access to health care is only a small part of the whole right to health. According to the World Health Organization, there are "underlying determinants of health", which are, of course, social determinants of health. They are the following:

- Safe drinking water and adequate sanitation
- Safe food
- Adequate nutrition and housing
- Healthy working and environmental conditions
- Health-related education and information
- Gender equality

There are many factors that can be identified with social determinants of health. First, we must take into account the "power of psychosocial factors: chronic anxiety, insecurity, low self esteem, social isolation, and lack of control over work appear to undermine mental and physical health." (Brunner and Marmot 2006: 28) These SDH are the 'causes of the causes' of disease. Nowadays, we can also see that "refugees and internationally displaced persons experience elevated risks of mortality in the period following their migration." (Shaw et al. 2006: 210) All these social determinants of health can be summed up by the fact that, for example, in the United Kingdom the difference in life expectancy between a rich neighbourhood of Glasgow and a poor one, in spite of the National Health Service, can be about 30 years (Puyol 2012: 179) There is something astonishingly unequal in the reality revealed by these figures.

Given the fact that citizenship is the means by which there can be respect for Human Rights, it is important to protect the right to health through democratic institutions. Democracy and social rights are necessary conditions for the development of a health care system and prevention programmes. However, as General Comment 14 of the Office of the High Commissioner for Human Rights states: "The Committee is aware that, for millions of people throughout the world, the full enjoyment of the right to health still remains a distant goal. Moreover, in many cases, especially for those living in poverty, this goal is becoming increasingly remote. The Committee recognizes the formidable structural and other obstacles resulting from international and other factors beyond the control of States that impede the full realization of article 12 (International Covenant on Economic, Social and Cultural Rights) in many States parties." (UNHCHR 2000: 2) The political level of States may not be enough in certain cases.

There is certainly a way to reverse this trend. Public policies have to take the SDH into account. This means that citizenship plays a crucial role in this field. Human Rights are respected through Citizenship Rights and solidarity. Citizenship Rights call upon the responsibility of the social institutions and the citizenship: "Participation of the population in health-related decision-making at the national and community levels." (WHO 2008: 4) The republican conception of citizenship is very relevant here because it establishes that citizens are not only passive consumers of goods and services. They are also active members of the community who can and must take part in the decision-making process. Public health is obviously a matter for experts. However, it is also true that social pressure and citizen initiatives often have an influence on political interventions. Through solidarity, citizens can also put pressure on political authorities to change their relationship with Third World countries and give them more help and all kinds of resources. In a word, responsible citizenship guarantees respect for the rights and continued exercises of international cooperation and solidarity.

The crucial issue is the qualified sense of 'public' in the term "Public Health". By the way, Republicanism (in the sense of 'common wealth' or the Latin expression 'res publica') is a theoretical perspective that casts a normative light on the SDH. From this normative dimension, it is very clear that political apathy and the lack of civic and active participation, in the long run, undermines citizenship rights (Dagger 1997) In the case of the 'right to health' this is especially important. Therefore, we end up with a political schema by virtue of which citizenship and social epidemiologists relate with each other. On the one hand, active citizenship puts pressure on the government to redistribute social determinants of health. On the other hand, social epidemiologists and public health planners try to change habits and transform the environmental conditions for citizens. These planned actions can be seen as 'educational' programmes for citizenship. There are a number of examples that show how these political measures can increase the level of population health: the tobacco case is paradigmatic of this trend (Jarvis and Wardle 2006: 224).

Public Health and epidemiology are the result of citizenship claims, concerns and demands. Republican Ethics of civic virtues can be actualized in order to show how all citizens must try to develop healthy habits (responsibility) and cooperate for the redistribution of SDH (solidarity). The republican interpretation of Rawls and Habermas highlights how citizenship rights depend on citizenship duties. Rawls calls them "duties of civility", one good example of which could be to take care of and cooperate with political institutions in order to prevent social determinants of health from being an obstacle for justice.

In order to secure citizenship rights and responsibilities, we need to take into account three kinds of State obligation that must be satisfied: namely, obligations to respect, to protect and to fulfil (WHO and UN HCHR 2008: 25). This formulation is very common in United Nations documents. In general, it is essential to consider the fact that States must ensure the right to access to a health system on a non-discriminatory basis: access to a minimum of essential food; access to housing, drinking water, sanitation and shelter; the provision of essential drugs; and the equitable distribution of health facilities (WHO and UN HCHR 2008: 25).

First, the obligation to respect implies refraining from interfering (directly or indirectly) with the right to health. States must respect this right not only inside their territory, but also in other countries. All discriminatory or limiting interventions, such as forbidding access to the health system to a particular part of the population (on the basis of gender or culture) or even infringing the right to privacy of patients, are breaches of the obligation to respect the right to health by the State (in this case, the obligation to respect is a *sufficientist* approach to the right to health).

Second, there is the obligation to protect. This obligation refers to the action of third parties. The State must "prevent third parties from interfering with the right to health." (WHO and UN HCHR 2008: 26) This means that the State should carry out legislative measures so as to secure access to health information and care. In terms of the SDH, this implies proactive action by the State to prevent private interests from being at odds with the right to health. According to the World Health Organization, privatization is one of the most important threats to the right to health. The obligation to protect the right to health goes further than securing an equal right for citizens to access the health care system. It is also related with the composition of food products, pollution and the natural environment, prevention of harmful traditional practices, information, and health education.

Third and finally, there is the obligation to fulfil, which refers to the full realization of the right to health. This stresses the administrative, legislative, budgetary, judicial, promotional and educative dimensions of the State. Although social determinants of health can be relevant in the obligation to respect and protect, it is in the obligation to fulfil that they appear to be unavoidable. It is not possible to fulfil the right to health without appropriate intervention regarding SDH. There is a need to "ensure equal access for all to the underlying determinants of health, such as safe and nutritious food, sanitation and clean water." (WHO and UN HCHR 2008: 27) The highest attainable standard of health can only be accomplished with effective SDH intervention. The document on the right to health does not include all SDH in its definition. However, the expression "underlying determinants of health" indicates some necessary conditions for health. Moreover, we should also include social services, job security, and the satisfaction of material and psychosocial needs.

Conclusion

Social determinants of health (SDH) produce health inequities and prevent the full realization of the right to health. One way of accomplishing the highest attainable standard of health is by means of political redistribution of SDH and active participation by the citizenship. The republican conception of citizenship stresses both rights and responsibilities in order to achieve social justice. Social justice, furthermore, is closely related to health. It is necessary to furnish a normative justification of redistributive political programmes and legislation, which can be formulated in terms of the principle of justice in bioethics (Beauchamp and Childress 2012). However, at the same time we should consider how there is also a need for a principle of responsibility. Active citizenship, in republican thought, means a form of "caring democracy" (Tronto 2013). Solidarity and responsibility are two sides of the same coin: citizenship engaged with the value of health equity and in defence of the right to health put pressure on governments to start to deploy more sensitive policies.

The redistribution of SDH is a praxiological action that affects at the same time the health level and the justice level of society as a whole. This just redistribution of SDH is a way of preventing diseases and is obviously a way of increasing population health. Preventing, from this point of view, is even more important than curing. Thus, public health becomes a normative matter in which the casuistic approach to bioethics and republican theories of citizenship cooperate to bring about guidelines for policy makers. Ethical studies of SDH are new examples of applied philosophy. Once the analyses of social epidemiology have tracked the social determinants of health, there will be a series of information ready to be used to reduce health inequities. Ethics, as a normative discipline, can help to make explicit the values that are at stake in such studies and interventions. Therefore, normative theories can be very useful to enable a more reflexive practice of science.

SDH are the causes of the causes of disease. The development of a bioethical approach to the problem of health inequities is blurring the boundaries between science and humanities, theory and practice, the descriptive and the normative, epidemiology and sociology, and politics and ethics. Epidemiology is becoming social epidemiology. And consequently, social epidemiology calls for political actions in particular areas of concern: poverty, labour conditions and gender discrimination, among others. In this sense, policy-makers and citizens should affect each other in terms of pressure and education. On the one hand, active citizenship puts pressure on the government to redistribute SDH and, on the other, public health planners contribute to an educated and responsible citizenship, whereby the right to health (World Health Organization) can be respected, protected and fulfilled as a Human Right. All in all, SDH research and application is fundamental for a secured and consolidated citizenship.

References

Aloisi, M., Berlinguer, G., Conti, L., Massucco, A., Misiti, R., Mondella, F., Omodeo, P., Piersanti, F., Seppilli, T., Somenzi, V., & Vegetti, M. (1968). *La medicina e la società contemporanea*. Roma: Editori Riuniti.

Beauchamp, T. L., & Childress, J. F. (2012). *Principles of biomedical ethics*. New York: Oxford University Press.

Berkman, L. F., & Kawachi, I. (2000). *Social epidemiology*. New York: Oxford University Press.

Broadbent, A. (2013). *Philosophy of epidemiology*. New York: Palgrave Macmillan.

Brunner, E., & Marmot, M. (2006). Social organization, stress, and health. In Marmot & Wilkinson (Eds.), *Social determinants of health: The solid facts*. Copenhagen: Centre for Urban Health, World Health Organization.

Cwikel, J. (2006). *Social epidemiology: Strategies for public health activism*. New York: Columbia University Press.

Dagger, R. (1997). *Civic virtues*. New York: Oxford University Press.

Daniels, N. (2008). *Just health*. New York: Cambridge University Press.

Daniels, N., Kennedy, B. P., & Kawachi, I. (2007). Why justice is good for our health: The social determinants of health inequalities. In R. Bayer et al. (Eds.), *Public health ethics: Theory, policy, and practice*. Oxford: Oxford University Press.

Gray, A. M. (1982). Inequalities in health. The black report: A summary and comment. *International Journal of Health Services, 12*(3), 349–380.

Habermas, J. (1991). *Moral consciousness and communicative action*. Cambridge: MIT Press.

Jarvis, M. J., & Wardle, J. (2006). Social patterning of individual health behaviours: The case of cigarette smocking. In Marmot & Wilkinson (Eds.), *Social determinants of health: The solid facts*. Copenhagen: Centre for Urban Health, World Health Organization.

Jonsen, A. R., Siegler, M., & Winslade, W. J. (2001). *Clinical ethics*. New York: McGraw-Hill.

Kotarbinski, T. (1965). *Praxiology. An introduction to the sciences of efficient action*. New York: Pergamon Press.
Marmot, M., & Wilkinson, R. (Eds.). (2006). *Social determinants of health: The solid facts*. Copenhagen: Centre for Urban Health, World Health Organization.
Marshall, T. H., & Bottomore, T. (1992). *Citizenship and social class*. London: Pluto Press.
O'Campo, P., & Dunn, J. R. (2012). *Rethinking social epidemiology: Towards a science of change*. Dordrecht: Springer.
Pettit, P. (1997). *Republicanism: A theory of freedom and government*. Oxford: Clarendon Press.
Potter, V. R. (1971). *Bioethics: Bridge to the future*. Upper Saddle River, NJ: Prentice Hall.
Puyol, A. (2012). Ética, equidad y determinantes sociales de la salud. *Gaceta sanitaria, 26*(2), 178–181.
Rawls, J. (1999). *A theory of justice*. Cambridge, MA: Harvard University Press.
Sen, A. (2002). Why health equity. *Health Economics, 11*, 659–666.
Shaw, M., Dorling, D., & Smith, G. D. (2006). Poverty, social exclusion, and minorities. In Marmot & Wilkinson (Eds.), *Social determinants of health: The solid facts*. Copenhagen: Centre for Urban Health, World Health Organization.
Toulmin, S., & Jonsen, A. R. (2009). *The abuse of casuistry: A history of moral reasoning*. Berkeley: University of California Press.
Tronto, J. (2013). *Caring democracy*. New York: New York University Press.
UN HCHR. (2000). *General Comment 14 The Right to the Highest Attainable Standard of Health* (art. 12).
Upshur, R. (2002). Principles for the justification of public health intervention. *Canadian Journal of Public Health, 93*(2), 101–103.
Venkatapuram, S., & Marmot, M. (2009). Epidemiology and social justice in the light of social determinants of health research. *Bioethics, 23*(2), 79–89.
Whitehead, M. (1990). *The concepts and principles of equity in health*. Copenhagen: World Health Organization Regional Office for Europe.
WHO, & UN HCHR. (2008). *The right to health*. (Fact sheet n° 31).
Winslow, C. E. A. (1920). The untilled fields of public health. *Modern Medicine, 2*, 183–191.

The Public Dimension of Care: Towards a Public Ethics of Care

Txetxu Ausín

*The place of other (*autrui*) is the true perspective point in politics as well as moral, and the precept of Jesus Christ, to put oneself in the place of other, serves not only the end of which our master spoke, namely morality, but that of politics as well.*

LEIBNIZ, *La place d'autrui, 1679 (A VI, 3, 903–904).*

Abstract As essentially social beings, we humans are fragile and dependent on what is outside ourselves—others, institutions and sustained and sustainable environments—. Vulnerability acquires for us a social role (not a mere contingency or a subjective disposition) and hence the positive social obligations arise to minimize instability and its differential distribution, to demand basic support (food, shelter, work, healthcare, education, mobility, expression) and to reduce avoidable damage. We are a social species, with reciprocal links (rights and duties), built on the interdependence; not a group of solitary individuals whose mutual obligations are limited to not invading other people's space. Care-related obligations by the (social) state come into play here, as in the case of healthcare systems. These can be characterized as "due cares" (due actions) because the action is an expected event within a social context of values and guidelines. In that context, the omission that causes or enables harm, as in exclusion from healthcare, qualifies as an "undue inaction". In the criminal legal sense, the idea of "nonfeasance" refers to bringing about a result by not avoiding it, by violating a special duty, which equates with its causation—the prototypical example is the crime of failing to assist. Here the question not only of intentional omission but of neglect and abandonment comes into play. One example is the exclusion from healthcare and drugs of wide sectors of

This work has been sponsored by the Spanish Ministry of Economy and Competitiveness through its research project KONTUZ! (FFI2014-53926-R): www.kontuz.weebly.com; INBOTS (EU-H2020 grant agreement 780073); and EXTEND (EU-H2020 grant agreement 779982).

T. Ausín (✉)
Institute of Philosophy, CSIC (Spanish National Research Council), Madrid, Spain
e-mail: txetxu.ausin@cchs.csic.es

J. Vallverdú et al. (eds.), *Philosophical and Methodological Debates in Public Health*, https://doi.org/10.1007/978-3-030-28626-2_10

the Spanish population, not only undocumented migrants but also poor and unemployed people. The State is shamelessly neglecting important segments of the population, and causing harm—especially to the most vulnerable—by an ethically reprehensible omission of care. This case illustrates the need for defining what we have called "a public ethics of care" combining public ethics obligations with the responsibilities of care—since the ethics of care shapes government.

Keywords Vulnerability · Care · Omission · Public health

Vulnerability: Existential and Uneven

The starting point of this reflection on care, justice and politics is the idea of vulnerability. To be vulnerable is to be capable of being wounded, receiving a damage or injury, to receive physical or moral injury, to suffer. The Latin term 'vulnus' means injury, beat, stab and misfortune or affliction. The vulnerability is related to the possibility of suffering, to the disease, to pain, to fragility, to the limitation, and finally to death.

In this sense, vulnerability is a basic constituent of the human being. Borges in *The Immortals* said: We are precious and pathetic beings. Among mortals, everything is precious, precarious, running every act may be the last.

Vulnerability motivates ethical concern for the fragility of the human condition. The human condition is marked by an extreme degree of fragility because of the temporal and finite character of all human life. The bodily incarnated human subject is destined for death and it is not possible to abstract it from mortality and destiny as basic to human life. Instead we have to live with mortality but also to take care of the vulnerable situated subject. So, we can say that vulnerability is actually our human condition.

We can find this kind of approach in different philosophers. Habermas, for example, stands his argument for communicative understanding in a domination-free dialogue and situates vulnerability in the center of the ethical concern. This is openness towards the vulnerable other being taking part in a dialogue. Lévinas has defined the concept of vulnerability as the foundation for understanding human condition. He analyses vulnerability as the foundation for morality. Morality is a compensation for human vulnerability. The moral imperative is an imperative to take care of the other and an ethical responsibility for the other. In this way vulnerability can be said to imply an immanent normativity in which the vulnerability is expressed in the corporeal incarnation of the other, e.g. in the face of the other. (About this "positive" sense of vulnerability will back after). Arendt, in her *Human Condition*, also says that human condition itself relays on fragility and vulnerability. Thus, we have an universal identity in pain, suffering and vulnerability finally.

However, vulnerability is also asymmetrical and uneven.

Some people or group are at disadvantage that puts them in a position of helplessness and susceptibility of injury. As pointed out by Iris Marion Young, any given society is comprised of a variety of social-structural positions that determine our more or less vulnerability (Young 2011, pp. 43–74).

According to Lévinas, vulnerability manifests an asymmetrical imbalance between the weak and the powerful and in this context it demands an ethical engagement and that the powerful protect the weak. It is our vulnerability that makes us receptive for the responsibility emanating from the other as a vulnerable being. Spanish philosopher Gabriel Bello (2011) stands that it is wrong to represent all as equally vulnerable or as equal in vulnerability, as symmetrically vulnerable. The basic fact is that human vulnerability is uneven or asymmetrical between each other, or that we relate to each other from our asymmetric vulnerability. Rather than assuming the fiction that all citizens are equal, a care perspective would have us recognize the achievement of equality as a political goal.

So, although we start from a primary, common vulnerability, there are situations, structures, processes, people and vulnerable groups[1] where damage, suffering, pain, abandonment or helplessness are the result of social and environmental elements. There is a variable and selective dimension of vulnerability that relies on factors that can be changed and on which it can intervene. That is what is called "realms of vulnerability": the set of unfavorable conditions that put people at higher risk situations, at lack of power or control, at the inability to change their circumstances and therefore at vulnerability. This is the "precarious life" to which Judith Butler relates: Life is cared for and maintained differentially and there are radically different distribution forms of vulnerability throughout the planet.

Therefore, vulnerability, even being intrinsic to human beings, is not a stable and unchanging but a dependent feature, at least in part, of factors that can be changed and which can intervene. While the intrinsic vulnerability can be reduced by equal protection of all members of society, social vulnerability (susceptibility) also requires specific active measures (repair, restoration).

In this context, vulnerability can also be interpreted as an important legal concept and even as the foundation for the legal system. The English philosopher of law and jurisprudence Hart, says that the vulnerability of human beings is the background for the regulation of its activities in rules and social institutions. The task of legal organisation, legal principles and concrete legal rules is to protect the vulnerable human being, confronted with the possibility of destruction and the interventions from other people and the state. For example, it is the task of legal regulations of biomedical problems to protect the weakest and the poorest in society against the discrimination and destruction from other social groups.

This asymmetry is clear, for example, about health. As pointed out by Venkatapuram (2011), vulnerability to impairment or premature mortality is a direct result of engaging in social cooperation (see chapter "Lost in Associations: How epidemiology's Epistemology Upholds Class, Race and Gender Inequities"). Health

[1]It should be notice that sometimes there has been a rutinary, non-reflective, and indeed incoherent use of concepts like "vulnerable groups". This concept could lead to stigmatization and stereotype of some populations.

outcomes are adversely affected by poverty, unemployment, and poor living conditions. Those factors beyond an individual's control undermine her ability to make good personal choices across all areas of her life, health included. Health is about more than healthcare; it is about education, income and the environment in which we live. A recent report in *The Lancet* about poverty and health clearly shows that low socioeconomic status is one of the strongest predictors of morbidity and premature mortality worldwide. It is the well-known concept of "social determinants of health". Health and illness are influenced by factors other than access to healthcare and the quality of care. Socioeconomic circumstances and their consequences are modifiable by policies at the local, national, and international levels, as are risk factors targeted by existing global health strategies.

> The poor health of the poor, the social gradient in health within countries, and the marked health inequities between countries are caused by the unequal distribution of power, income, goods, and services, globally and nationally, the consequent unfairness in the immediate, visible circumstances of people's lives—their access to health care, schools, and education, their conditions of work and leisure, their homes, communities, towns, cities—and their chances of leading a flourishing life. This unequal distribution of health-damaging experiences is in no sense a 'natural' phenomenon but it the result of a toxic combination of poor social policies and programs, unfair economic agreements, and bad politics. (Commission on Social Determinants of Health 2008, p. 1)

Care: Empathy and Responsibility

To the extent that the human being is inevitably open to injury, it is also open to cure-care. The vulnerability is in this tension: cure/care. Cure and care are inextricably linked. The Latin verb 'curo' means caring, worrying about. And the term 'cure' (treatment) has this sense of care, solicitude, care, treatment. To prevent damage, to minimize, mitigate or prevent harm conditions (realms of vulnerability), we must be careful, we must care: to care for, to care about, and to take care of (involving some kind of engagement). So, vulnerability is therefore an extreme important concept as the foundation for ethical notions of care and responsibility. Thus, vulnerability is not only a negative human feature but a positive one since promotes cooperation, solidarity, assistance and care.

Human life it would be inconceivable without relations of care. So, there is a universality in caring too. Care is understood as one of the central activities of human life and the ethics of care has put at the center of ethical reflection precisely the idea of care. A wide accepted definition of care is the following:

> We suggest that caring we viewed as a species activity that includes everything that we do to maintain, continue, and repair our "world" so that we can live in it as well as possible. That world includes our bodies, ourselves, and our environment, all of which we seek to interweave in a complex, life-sustaining web. (Fisher and Tronto 1990, p. 40)

This definition posits care fundamentally as a practice, but Tronto further identifies four sub-elements of care that can be understood simultaneously as stages, virtuous dispositions, or goals. These sub-elements are: (1) attentiveness (caring about/empathy), a proclivity to become aware of need; (2) responsibility (caring for), a willingness to respond and take care of need[2]; (3) competence (care giving), the skill of providing good and successful care; and (4) responsiveness (care receiving), consideration of the position of others as they see it and recognition of the potential for abuse in care (Tronto 1993, pp. 126–136).

Care is about the social reproduction, that is, on the creation and maintenance of social bonds. One part of this has to do with the ties between the generations—so, birthing and raising children and caring for the elderly. Another part is about sustaining horizontal ties among friends, family, neighbourhoods, and community. This sort of activity is absolutely essential to society; without it, there would be no social organization, nor economy, nor culture, nor politics...

It could be connect this approach to the Foucault's concept of "care of the self" (Foucault 1986), interpreted as the ability to see oneself as vulnerable. We are all agents and receivers of care, we need care as vulnerable and interdependent beings, we care of and are cared for in a virtuous circle. It is not an "other", it is an "us". Nevertheless, there is a chronic deficit of recognition of care tasks.

The perspective of care clearly rejects the individualistic moral model of Kantian inspiration, which has in turn determined the visions of moral development à la Kolhberg (Baier 1988).

We can summarize the values and practices associated with the ethics of care as follows: pay attention to others, attention to context, sensitivity to each other's needs, emphasis on human vulnerability and dependence, and relational understanding of self.

Interdependence

We are a social species, with reciprocal links (rights and duties), built on the interdependence, not a group of solitary individuals whose mutual obligations are limited to not invading other's people space.[3] The vulnerability [fragility] of human beings involves accepting a relational anthropology against modern individualism and unencumbered, sovereign and independent self. The recognition of human vulnerability is a critique of the myth of an independent and disembodied subject, a subject that is not born, or sick, or age, or lose power.[4] As Tronto says, we will

[2]Empathy and responsibility are the main features of the ethics of care. The discovery of mirror neurons shows that empathy is a fundamental human capacity that we are born with but which must be strenghtened through a nurturant upbringing or it will decay. But, as Lakoff stands, caring is not just feeling empathy; it is taking responsibility, acting powerfully and courageously. You have to be strong to care, and to act on that care with success; you should be responsible about care (Lakoff 2009, p. 47).

[3]This idea of "extended self" belongs to philosphers as Ortega y Gasset or Heidegger.

[4]"Throughout our lives, all of us go through varying degrees of dependence and independence, of autonomy and vulnerability. A political order that presumes only independence and autonomy as

need to rethink our conceptions of human nature to shift from the dilemma of autonomy or dependency to a more sophisticated sense of human interdependence (Tronto 1993, p. 101).

Thus, interdependence would be the state that best characterizes us as members of a social community: Interdependence is a value derived from human relationships themselves. No human being is, or has been, one, independent, sovereign, self-determined, etc. We depend strictly on others, and not only on our family when we are small, but throughout our whole life. Instead of thinking of ourselves as sovereign and autonomous, we must take into account how many forms of heteronomy we accept daily. Being the cooperation, interdependence and plurality of constitutive values of human relations, it would be worthwhile to project these values to social relations in general, affirming them as fundamental.

Even autonomy should not be interpreted as independence because even to develop and put into practice our autonomy, we need others—relational autonomy or autonomy with others in the sense of feminism (Mackenzie and Stojlar 2000). At the end of radical individualism is radical authoritarianism. The example is Donald Trump, the exacerbation of an individualistic logic characteristic of some contemporary societies.

As essentially social beings, we humans are fragil and dependent on what is outside ourselves: others, institutions, and sustained and sustainable environments. People do not spring up from the soil like mushrooms. People produce people. People need to be cared for and nurtured throughout their lives by other people, at some times more urgently and more completely than at other times (children, elderly, dementia, . . .). Thus, identity is extended to include the experience of interconnection, so that the moral domain is widened by the inclusion of responsibility and care.

Due Care and Omissions

Vulnerability acquires for us a social role (not a mere contingency nor a subjective disposition) and hence the positive social obligations arise to minimize instability and its differential distribution, to demand basic support (food, shelter, work, healthcare, education, mobility, expression) and to reduce avoidable damage. Living in society is to care each others. As said Obama, the core of patriotism is caring for each other and, I add, not to do not wave flags or discuss national anthems as in Spain today.

Responsibility is a central topos in the ethics of care. There are care-related obligations (due actions) because some action is an expected event within a social context of values and guidelines. These expectations of care (necessary for social reproduction) are not limited to an intimate or family sphere, based on the ideology

the nature of human life thereby misses a great deal of human experience, and must somehow hide this point elsewhere". (Tronto 1993, p. 135).

of the naturalization of feelings of commitment (supposed feminine nature), such as charitable, stigmatizing and voluntarist actions, without any social responsibility beyond individual or family responsibility. It should be overcome that traditional conception of care as housework, family, intimate, and non-productive.

> To recognize the value of care calls into question the structure of values in our society. Care is not a parochial concern of women, a type of secondary moral question, or the work of the least well off in society. Care is central concern of human life. It is time to began to change our political and social institutions to reflect this truth. (Tronto 1993, p. 180)

Neoliberalism frames care as a service commodity: Use the market to care for yourself and your family. In fact, caring needs are heavily assigned to less powerful people, marked along lines of gender, class, and ethnicity by the buying in the market of caring services to less powerful people (women, migrants).[5] The way in which caring responsibilities are allocated in society leads to 'privileged irresponsibility': the phenomenon that allows the most advantaged in society to purchase caring services, delegate the work of care-giving, and avoid responsibility for the adequacy of hands-on care.[6]

> We now have a dual organization of care work in which those who can afford domestic help simply pay for it, while those who cannot scramble to take care of their families, often by doing the paid care work for the first group, and often at very, very low wages with virtually no protections. (Fraser 2016, p. 34)

Private care is reinforced by the separation of the public and private spheres where care is devalued, connected to privacy, emotion, and needy people.

> (...) Because socially privileged groups can most easily purchase alternative private sources (clean water for example) they have the least interest in maintaining in generally good condition collective goods and services. (Plumwood 2002, p. 85)

Moreover, there is a tendency to the total mechanization of care (technological solutionism) but activities oriented to sustaining social bonds contain an ineliminable personal element because they are, by definition, interpersonal, involving intersubjective communication and in some cases physical touching. Care, for example in medicine, cannot be reduced to technical competencies instead of seeing the relational dimension of care as a point of departure. In fact, the best healthcare innovation is quality universal care access. All those wearables and techs are distractions if they exist in unequal scenario.

Thus, the duties of care have a social, public, and institutional dimension; responsibility has a social and public dimension not only individual. In that context, the omission that causes or enables harm qualifies as an "undue inaction".

[5]Outsourcing care work creates a "global care chain" (Hochschild and Ehrenreich 2003).

[6]"New forms of capitalism try to insert care into the market, and increasing areas of activities that used to be in the public sphere are being privatized. Modern analysis has led to the emergence of a kind of neoliberal governance in respect of care through the extension of the market into social and personal spheres." (Brugère 2011, p. 73).

Remember that some social and political structures allow certain groups in society to elude their responsibility (privileged irresponsibility).

The main concern of the ethics of care was the harm caused by relationships that lack care (Gilligan 1982). It is harm caused by the absence of active beneficent participation in the relationship: by not caring, we may hurt the other. So, the focus of the ethics of care is this 'relational' harm:

> Instead of focusing on the danger of interference in terms of violating a person's autonomy, an ethics of care put focus on the harm caused by abandonment and a lack of care in the relationship. (Martinsen 2014, p. 122)

Here the question not only of intentional omission but of neglect and abandonment comes into play—lack of prevention and precaution: Non-abandonment is an important moral obligation in medicine. That is, harm can be caused when care is absent.[7]

The collective organization of care relies on institutions and caring organizations that often fail to meet the needs of those to whom they pretend to be dedicated. In the criminal legal sense, the idea of "nonfeasance" refers to bringing about a result by not avoiding it, by violating a special duty, which equates with its causation[8]—the prototypical example is the crime of failing to assist. (I'm aware of the difficulties on the role of causation by omission).

Two Cases of Undue Inactions

Two prominent cases of failing to assist in healthcare in Spain are described briefly in following:

5.1. The exclusion from healthcare of wide sectors of the Spanish population, not only undocumented migrants but also poor and unemployed people. The State is shamelessly neglecting important segments of the population, and causing harm—especially to the most vulnerable—by an ethically reprehensible omission of care. One of the greatest achievements of democracy in Spain has been the creation and consolidation of a public health care system that implemented

[7]Julian Tudor Hart espoused the "inverse care law": As much more care is needed, less care received, and moreover in a market oriented health system.

[8]I'm aware of the difficulties on the role of causation by omission. About this topic, see the research project "Causal responsibility of perpetration through nonfeasance: An ethico-juridical inquiry into undue inaction" (www.kontuz.weebly.com). *We seek to determine the causal responsibility of nonfeasance and the problems posed thus by undue inaction. To this end, we will combine a social action approach to action with a materialistic conception of causality, a factualist ontology that introduces nonexistence or negative facts, and a gradualist theory that will accommodate degrees of causation, degrees of social expectation, and of ethical-legal responsibility, all in line with the earlier work done by this research group and with the introduction by cognitive scientists and psychologists of moral categories such as "enabling" harm* (Hanna 2015).

the right to health protection enshrined in the 1978 Spanish Constitution (Art. 43). The subsequent legislative development progressively secured the right to healthcare for all inhabitants residing in the country, without exception. Thanks to this formulation, the Spanish National Health System has significantly contributed to the achievement of high levels of social cohesion. Such a system has been possible mainly because the full funding of the health care system passed to be in charge of the General State Budgets, meaning that health care services were decoupled from individual contributions to the Social Security. The right to health protection was thus configured as a right of citizenship. However, this situation has changed since the entry into force, in 2012, of a Decree: Royal Decree-Law 16/2012, *on urgent measures to ensure the sustainability of the National Health Care System and enhance the quality and safety of its service* [RDL16], approved on April 20th, after a brief debate in the Congress and thanks to the parliamentary majority of the conservatives. According to this law, undocumented migrants will not be covered by the public healthcare system, free of charges, unless they are minors, pregnant women, are in an emergency situation or have contagious diseases. It is important to note that there are also other population groups with Spanish nationality who have been excluded from the health care system, such as those who have resided more than 3 months abroad or have been long-term unemployed. The modifications introduced by the RDL16 have involved not only a serious restriction of undocumented population's access to health care, but also a radical paradigm shift in the essence of the Spanish health system, a model originally based on the pillars of solidarity, equity and accessibility. In practice, the law implies that health care is no longer public, universal and free for all residents of the country, but only for those who are insured (and their dependents or beneficiaries). The exclusion from healthcare is a profound shift of paradigm regarding the right to health care in Spain which goes against many statements and declarations on Human Rights signed by our country. The outcome of the reform is not only to commit in a deep and flagrant injustice, but also to create a health system that is ineffective and inefficient: part of the population is excluded of the conventional health system; increasing in costs due to the use of emergency services; proliferation of cheating practices.[9]

5.2. Non-prescription of drugs needed for chronic diseases as hepatitis C. In 2017 two officials of the Spanish health administration were charged in a reckless homicide case because of delays in providing a drug for hepatitis C that could have caused at least six deaths. This is a consequence of cuts, austerity policies, and retrenchment in spending of social protection in Spain, as in other European countries. Some of the effects of austerity were observed in cuts to budgets for the public financing of health care, leading to increases in user charges for health services and prescription drugs while wage cuts and rising

[9]You can find a deep study on this topic in Triviño et al. (2016).

unemployment where making these charges more difficult to pay for those affected (Schrecker and Bambra 2015, p. 74).

A Public Ethics of Care

The concept of care is correlative to the dignity, integrity and vulnerability of the human person. Expectations of care involve a public and institutional dimension of care. That is the moral mission of government: caring about each other and acting responsibly on that care. So, care requires that government have two intertwined roles: protection (security) and empowerment.

Protection can be understood not only as police, army, fire department or emergency but also means social security, public health, safe food and water, labor and consumer protection, disaster relief or environmental protection. Even protection in relation to the power of the government and the administration (checks and balances) for which elements of good governance such as transparency, openness, accountability or participation are indispensable.

Empowerment consists of maximizing the freedom to achieve the goals and life projects of citizens. It includes actions such as communications, roads and bridges, public education, financial intermediation or the legal system and refers to professionalism, efficiency and innovation in public policies as necessary facilitators of personal fulfillment and sustainable human development of the community.

Duties of care, care required, due care, care expected, expectations of care are duties, commitments of "solidarity", which is an emerging concept in bioethics (Nuffield Council on Bioethics 2011). In this sense, care drives at organized solidarity (fraternity) with the collaboration of governments, institutions, and civil society organizations as NGOs. To the extent that care has an institutionally organized dimension and, therefore, it is necessary to determine the duties and institutional rules of care, we are faced with a question of justice. So, there is no opposition between the ethics of care and ethics of justice. There can be no justice without care nor care without justice.

Thus, fraternity, understood as the ideal of individuals sharing the burdens and benefits of the community to which they belong, would be a necessary pre-requisite for overcoming the deficits of justice generated by the neutrality of the state and the impartiality demanded by the liberal approach to justice (Puyol 2017, p. 25).

Employing expanded ideals of fairness and reciprocity that take interdependence as basic, Kittay poses a third principle for Rawls' theory of justice (a principle of the social responsibility for care):

> To each according to his or her need for care, from each to his or her capacity for care, and such support from social institutions as to make available resources and opportunities to those providing care. (Kittay 1999, p. 113)

The ethics of care shapes government and thus, care is a key concept of ethical and political theory that includes societal structures as the subject matter of critique and transformation.

. It is called "Public Ethics" to refer to applied ethics relating to the affairs of government and state organizations and institutions. It is, therefore, an ethics applied to public servants as agents of a panoply of joint activities with a view to favoring the common good, that is, the satisfaction of an ordered set of social preferences related to well-being, care, and service to the population in terms of equity. And because of this, Public Ethics is nothing more than an Ethics of Care that addresses the unequal distribution of vulnerability in society and which takes into consideration the expectations of care of individuals and groups.

From this "Public Ethics of Care" approach, justice will seek to protect the worst-off and vulnerable and transform the bases so that they stop to be so, placing care at the center of public policies. This goal is crucial in the current context of a capitalism's crisis of care as Fraser says: When a society simultaneously withdraws public support for social reproduction (care) and conscripts the chief providers of it into long and grueling hours of paid work,[10] it depletes the very social capacities on which it depends. This is exactly our situation today. The current, financialized form of capitalism is systematically consuming our capacities to sustain social bonds, like a tiger that eats its own tail (Fraser 2016, p. 31).

Thus, there is no justice without care nor caring democracy without democratizing caring. In her last paragraph of *Caring Democracy*, Joan Tronto stands:

> There is a way to turn our world around. It requires us to recommit to caring for ourselves and others by accepting and rethinking our caring responsibilities and providing sufficient resources for care. If we are able to do this, then we will able to enhance levels of trust, reduce levels of inequality, and provide real freedom for all. In such a society, we would have to conclude, with Giambattista Vico (1990[1709], 67): What is justice? It is the constant care for the common good. (Tronto 2013, p. 182)

References

Baier, A. (1988). The need for more than justice. *Canadian Journal of Philosophy, 13*. (M. Hanen & K. Nielsen (Eds.), pp. 41–56).

Bello, G. (2011). *Emigración y ética. Humanizar y deshumanizar*. Madrid: Plaza y Valdés.

Brugère, F. (2011). *L'éthique du "care"*. París: Presses Universitaires de France – PUF.

Commission on Social Determinants of Health. (2008). *Closing the gap in a generation: Health equity through action on the social determinants of health (Final report)*. Geneva: WHO.

Fisher, B., & Tronto, J. C. (1990). Toward a feminist theory of caring. In E. K. Abel & M. Nelson (Eds.), *Circles of care*. Albany: SUNY Press.

Foucault, M. (1986). *The history of sexuality. Vol. 3: The care of the self*. New York: Vintage.

Fraser, N. (2016, Fall). Capitalism's crisis of care. *Dissent*. Accessed November 23, 2017, from https://www.dissentmagazine.org/article/nancy-fraser-interview-capitalism-crisis-of-care

Gilligan, C. (1982). *In a different voice: Psychological theory and women's development*. Cambridge, MA: Harvard University Press.

[10](. . .) as the caregiver has the responsibility to care for the dependent, the larger society has an obligation to attend to the well-being of the caregiver. Only so can the caregiver fulfill responsibilities to the dependent without being subject to an exploitation some have called 'compulsory altruism' (Kittay 1999, p. 133).

Hanna, J. (2015). Enabling harm, doing harm, and undoing one's own behavior. *Ethics, 126*(1), 68–90.
Hochschild, A. R., & Ehrenreich, B. (Eds.). (2003). *Global woman: Nannies, maids, and sex workers in the new economy*. New York: Metropolitan Books.
Kittay, E. F. (1999). *Love's labor: Essays on women, equality, and dependency*. New York: Routledge.
Lakoff, G. (2009). *The political mind: A cognitive scientist's guide to your brain and its politics*. New York: Penguin.
Mackenzie, C., & Stojlar, N. (Eds.). (2000). *Relational autonomy: Feminist perspective on autonomy, agency, and the social self*. Oxford: Oxford University Press.
Martinsen, E. H. (2014). Caring in medicine: From a gentleman's care to a more sophisticated sense of human interdependence. In G. Olthuis et al. (Eds.), *Moral boundaries redrawn. The significance of Joan Tronto's argument for political theory, professional ethics, and care as practice*. Leuven: Peeters Publishers.
Nuffield Council of Bioethics. (2011). *Solidarity: Reflections on an emerging concept in bioethics*. Accessed January 14, 2018, from http://nuffieldbioethics.org/project/solidarity
Plumwood, V. (2002). *Environmental culture: The ecological crisis of reason*. New York: Routledge.
Puyol, À. (2017). *El derecho a la fraternidad*. Madrid: Los Libros de la Catarata.
Schrecker, T., & Bambra, C. (2015). *How politics makes us sick. Neoliberal epidemics*. New York: Palgrave Macmillan.
Triviño, R., Rodríguez-Arias, D., & Ausín, T. (2016). Exclusion from healthcare in Spain: The responsibility for omission of due care. In H. P. Gaisbauer et al. (Eds.), *Ethical issues in poverty alleviation*. Dordrecht: Springer.
Tronto, J. C. (1993). *Moral boundaries: A political argument for an ethic of care*. New York: Routledge.
Tronto, J. C. (2013). *Caring democracy: Markets, equality and justice*. New York: University Press.
Venkatapuram, S. (2011). *Health justice: An argument from the capabilities approach*. New York: Polity Press.
Young, I. M. (2011). *Responsibility for justice*. Oxford: Oxford University Press.

Genetic Prevention, Carrier Screening and the Role of Citizens: The Legitimisation of Libertarian Paternalist Policies by Deliberative Democracy

Baroukh Maurice Assael and Giovanni Boniolo

Abstract In this paper, first we recall some paradigmatic ways of implementing carrier screening policies in a community. Then we move on to a proposal concerning a possible bottom-up solution based on deliberative democracy (and thus on real citizens' participation in decisional processes) that should legitimate libertarian paternalist strategies concerning which genetic screenings should be made and which policies should be adopted by the institutions.

Keywords Genetic prevention · Carrier screening · Libertarian paternalism · Deliberative democracy · Citizen participation

Introduction

There is certainly no need to recall the importance of carrier screenings for genetic diseases and how much they can influence individual and collective choices, in particular reproductive choices. An open question however is whether the decision to undergo a screening test should only be individual or if a screening strategy to prevent genetic diseases should be a matter of collective decisions. In an era when genetic tests are more and more offered to "consumers" and given the high social burden of genetic diseases in contrast with the continuous decrease of the costs of testing, we feel that this needs to be discussed. The focus of what follows is, indeed, how a decisional process should be structured. Should a genetic screening strategy be planned by institutional health systems or is there a role for citizen participation? The question is relevant since the time of eugenics, when Francis Galton and other

B. M. Assael (✉)
Cystic Fibrosis Adult Center, University of Milano, Medical School, Milano, Italy
e-mail: benny.assael@fastwebnet.it

G. Boniolo
Dipartimento di Scienze Biomediche e Chirurgico Specialistiche, Università di Ferrara, Ferrara, Italy
e-mail: giovanni.boniolo@unife.it

J. Vallverdú et al. (eds.), *Philosophical and Methodological Debates in Public Health*, https://doi.org/10.1007/978-3-030-28626-2_11

biologists claimed that the state should intervene to ameliorate the population's genetic background.

Just to enter immediately *medias res*, we would like to mention recent studies that have shown how the incidence of cystic fibrosis (CF) has drastically declined in areas where the test to detect healthy carriers has been made available to the general population, as it happened, for example, in North Eastern Italy (Castellani et al. 2016) and in Israel (Stafler et al. 2016). This follows the decline and almost disappearance of other two genetically recessive severe diseases, i.e. thalassemia, once endemic in some Mediterranean areas notably in Cyprus and various Italian regions (Cao and Kan 2013; Angastiniotis and Hadjiminas 1981; Bozkurt 2007), and Tay-Sachs disease, highly incident among Jews of Ashkenazi origin (Kaback 2001).

The success of the primary genetic prevention and the availability of tests, especially at low costs, which can be offered to the general population may have great impact on the strategies to prevent a number of other monogenic recessive diseases,. Needless to recall, health policies promoting the diffusion of genetic tests to identify healthy carriers of recessive genetic diseases have raised, and raises, ethical debates about issues such as social stigmatization, right not to know, positive and negative eugenics, prevention through preimplantation diagnosis, reproductive choices, etc. But they also raise concerns about which genetic diseases should deserve a health policy strategy; on which ground they should be chosen (their social cost, their severity, their incidence?); and who should decide (the state?, a particular community?, the individual?).

In 2009, A. E. Raz published a comment on *Dor Yeshorim*: the premarital carrier screening program established among an orthodox Ashkenazi Jewish community. Beyond the interest of the sociological description of the program and its intended and unintended consequences, Raz posed a question: "Can population-based carrier screening be left to the community?". Such a question was focused on that particular case but it is worth rethinking it in a more general way rephrasing and completing it as "*Can population-based carrier screening be left to the citizens and in which way is this feasible?*". This is the very point we want to face, and it deals with the structure and the functioning of contemporary democratic societies, where the citizens should have an autonomous decisional power especially in health-related issues.

Before coping with that problem, we start by recalling some paradigmatic ways of implementing carrier screening policies in a community. As we will briefly report, some are institutional or religious top-down strategies, others are patient-spurred strategies, some are ethnically centred, while others panethnic. Beyond these planned strategies, it deserves to mention what, in a totally unplanned manner, happened in Italy with cystic fibrosis. By the way, this is, as we will see, a wonderful example of what the methodological individualists would call "unintended collective consequences of intentional individual human actions", or "order without a plan" (cf. Infantino 1998). We will shortly review some paradigmatic case studies of policies aimed at implementing carrier screening strategies and then move on to our theoretical proposal concerning a possible bottom-up solution based on deliberative democracy, which could legitimate libertarian paternalist policies.

Libertarian paternalism is a decisional approach based on the assumptions that (i) it is false that individuals always (or usually) make choices that are in their best interest; (ii) in many situations, some authority (i.e. the state) must make a (more or less) mandatory policy that will affect individuals' choices and life styles; (iii) not always paternalism involves coercion. Thus, according to libertarian paternalism, an authority should introduce *positive rules* related to proper lifestyles or health choices. Of course, these positive rules should be limited to decisions that are difficult, complex and infrequent, and when individuals have poor feedbacks and few opportunities for knowing and learning. In such cases an authority (the state) has the duty/right to intervene (Thaler and Sunstein 1993, 2003; Sunstein and Thaler 2003). Such libertarian paternalism is a variant of the so-called weak paternalism (an authority intervenes on ground of beneficence or non-maleficence only to prevent substantially bad consequences of actions made by individuals without a proper capacity of analysis, without adequate knowledge or information, or under irrational or false motivations). This means that the authority should not brutally force individuals to act in the way it thinks to be the best one, but it should study and provide *nudges* to "encourage" individuals to act in a certain way. Said differently, such an authority should be a sort of *choice architect* who designs apt decisional scenarios, where nudges have a preeminent role, so that individuals choose better (Thaler and Sunstein 2008).

The libertarian paternalism approach is rising a great interest in many countries and it has been the conceptual basis for many institutional campaigns regarding correct lifestyles and the implementation of health policies. For example, concerning tobacco, it has been proposed to sell it only to adults who have purchased an annual personal permit (this is the nudge); concerning salt in food, it has been proposed to produce unsalted foods (this is the nudge) so that consumers must deliberately add salt if they wish; concerning correct nutrition, it has been proposed an agreement with supermarkets for the prominent display of "healthy" foods and visible warnings for those that are potentially harmful, free distribution of fresh fruit in primary schools and the involvement of firms in similar schemes for their employees (these are the nudges).

By *deliberation* is meant any system of political (or ethical) decision based on some trade-off of consensus decision-making and representative democracy. In contrast to the traditional theory of democracy, which emphasizes voting as the central institution, deliberative democracy theorists argue that legitimate policies can only arise from the public deliberation of the citizenry. Very briefly, deliberation is the procedure through which people, starting from different initial positions, try to reach a shared decision *via* a debate based on rational arguments and counter-arguments (Boniolo 2012; Boniolo and Schiavone 2015). This deliberative approach has been used many times in the field of political and ethical public decisions concerning health and biomedical research policies. For example, it was used by the Health Services Commission of Oregon to involve citizens to decide which health plan should be adopted (Bodenheimer 1997), or by the Human Fertilisation and Embryology Authority of UK to decide whether the research on cybrids (cytoplasmic hybrids) was ethically plausible (Camporesi and Boniolo 2008; Taddeo and Robert 2014), and in many other cases (see Abelson et al. 2013; Boniolo and Di Fiore 2010; Boniolo et al. 2014, 2015; Boniolo and Rebba 2015; Mosconi et al. 2015, 2016).

Case Studies of Population Carrier Screening

State and Religion

Thalassemia is an inherited autosomal recessive disease[1] of the red blood cells. It affects the synthesis of haemoglobin and causes mild to severe anaemia. There are two main types of thalassemias: the alpha-thalassemias, which involve the genes HBA1 (Hemoglobin, alpha 1) and HBA2 (Hemoglobin, alpha 2), and the beta-thalassemias dealing with the haemoglobin beta gene (HBB).

The first campaigns for the prevention of thalassemia were conducted in endemic areas of the Mediterranean, such as in Italy and in Cyprus, in the middle of the past century. But in the two different locations, the related campaigns had different implementation strategies.

In Italy, the disease had a particular geographic distribution with endemic areas that reflect the historical presence of malaria, i.e. the south-western Veneto, Lazio, Puglia, Sicily and Sardinia. In these regions dissemination campaigns on carrier testing were conducted starting from the 1950s. Shortly, promoted by scientists, an awareness program was implemented among local public health officials of possible health individuals carrying the recessive disease. Then the Ministry of Public Health instituted some centres for microcythemia, which had, among their tasks, the mandate to diffuse the carrier test. In particular, testing was offered to the general population in high schools and during pregnancy. Although this was not a national coercive campaign, the support of central health authorities strongly encouraged individual choices and determined the success of prevention in areas where the disease was endemic (Cao et al. 1989; Canali and Corbellini 2002).

It is important to note that Italy was (and for certain aspects, still is) a country deeply rooted in Catholicism. This means, especially in the years of the thalassemia campaign, that the Catholic Church plays a strong role both for moral suasion or for moral dissuasion. And this happened both at local levels, i.e. at parishes level, and at general level, i.e. at level concerning hard and soft laws promulgated respectively by the Italian parliament or by governmental institutions but always with a strong Catholic orientation. It is to say that the Catholic Church did not explicitly participate in the campaign for the prevention of thalassemia, but it did not even hinder it. Actually, there was a move from the Vatican, in particular from the Pope, Pius XII. In 1959, he expressed himself in favour of the prevention of serious genetic diseases, particularly thalassemia, supporting the choice of abstinence or of the birth

[1]Even if some cases of dominantly inherited alpha and beta thalassemias have been reported.

control methods allowed by the Church, but, of course, banning abortion as possible reproductive choice.[2]

In Cyprus things went differently, even if it was the first country to get rid of thalassemia through a national campaign, started in the early 1970s. Here, the blood test, which was used to identify carriers, was supported both by the health authorities and by the Orthodox church that made premarital testing mandatory (raising for this reason some ethical concerns), leaving however the marriage and the reproductive choice to the couple (cf. Beck 1999). The Orthodox Church in Cyprus and the Catholic Church in Italy had different roles. The Catholic Church and the Italian state were formally independent, even if strongly intertwined by the 1929 Lateran Treaty, signed by the then King of Italy (Victor Emmanuel III) and the Pope Pius IX, under the auspices of Mussolini and of the fascist government, while in Cyprus, when the campaign, the President of the Republic was Makarios III, Archbishop and Primate of the autocephalous Orthodox Church of Cyprus.

Summing up, in Italy we had a top-down institutional health carrier campaign, characterised by being non-mandatory but strongly suggested. Moreover, it had a religious approval, of course within the consequent moral religious constrains. In Cyprus the top-down policy was characterised by a mix of institutional and religious aspects, the premarital test was mandatory, although formally liberal[3] concerning marriage and reproductive choices.

We should also mention that a screening campaign was also implemented in the Turkish part of Cyprus by the civil authority. Moreover, the Italian and Cyprus examples influenced the policies of many several Muslim countries that event today conduct genetic prevention policies, offering premarital counselling especially aimed to reduce marriages between first cousins. Actually here many different strategies has been implemented, some of which are mandatory, while some others optional but suggested, and with a different range of allowed reproductive choices[4] (Al-Gazali et al. 2006; Tadmouri et al. 2009; Hamamy 2013; Khorasani et al. 2008; Samavat and Modell 2004).

Patient Organisations

Spinal muscular atrophy (SMA) is a motor neuron disease characterized by the progressive degeneration of nerve cells in the spinal cord and brainstem, leading to muscle weakness, muscle atrophy, and respiratory complications. Among genetic diseases, it is

[2]See, http://w2.vatican.va/content/pius-xii/fr/speeches/1958/documents/hf_p-xii_spe_19580912_ematologia.html

[3]We say it was formally liberal, since there was no legal prohibition to marriage. Actually we do not know if it was also substantially liberal, or if there were some forms of social or familial prohibitions.

[4]For example in Iran even the possibility of abortion before the 16th weeks of gestation is admitted.

a leading cause of death among children under age two. SMA is inherited as an autosomal recessive trait. Molecular genetic testing has revealed that all types of autosomal recessive SMA are caused by mutations in the SMN (survival motor neuron) gene on chromosome 5. In 2004, Deborah and Chris Heine established *The Claire Heine Foundation* after the death of their daughter Claire caused by SMA. This foundation is a private organisation "to implement public health policy to promote pan-ethnic carrier screening for Spinal Muscular Atrophy" (http://clairealtmanheinefoundation.org).

This shows that patient organizations may have a role on strategies aiming at the prevention of specific diseases. Nevertheless, typically, their initiatives are limited and do not involve other diseases.

Another example is *The Canavan Foundation* which was "launched in 1992, when Morgan Gelblum, daughter of Orren Alperstein and Seth Gelblum, was 2 years old and recently diagnosed with Canavan Disease" (http://www.canavanfoundation.org). The Canavan disease (also called Canavan–van Bogaert–Bertrand disease or Spongy degeneration of the central nervous system) is an autosomal recessive genetic disease, due to mutations in the ASPA gene, that provides instructions for making an enzyme called aspartoacylase. This causes deterioration of the white matter (myelin) in the brain, thereby preventing the proper transmission of nerve signals, that is, it causes progressive damage to nerve cells in the brain, and is one of the most common degenerative cerebral diseases of infancy.

Actually, there are two major differences between *The Claire Heine Foundation* and *The Canavan Foundation*. The former is pan-ethnically oriented, while the latter is ethnically oriented. Moreover the former is focused only on a particular disease, i.e. the SMA; the latter is "dedicated to educating at-risk populations about Canavan disease and other Jewish genetic diseases", although Canavan disease may also affect other ethnic groups (specifically, Saudi Arabians) but with a different incidence: in non-Ashkenazi about 1 out 100,000 births; instead if both parents are of Ashkenazi Jewish descents, about from 1 out 6400 to 1 out 13,500 (http://ghr.nlm.nih.gov/condition/canavan-disease; Matalon and Michals-Matalon 1993; Haldeman-Englert 2013).

So far we have considered examples where a group of patients or patients' relatives decided to organise themselves in order to rise awareness about a specific disease and possibly to make pressure on the institutions to make them implement a general policy. This gave rise to both ethnic and panethnic screening programs. Ethnically oriented policies have some peculiarities. They may involve a strongly cohesive community on a social or religious basis. They are more cost effective, since investigations are promoted in groups highly affected by the disease, moreover they are usually geographically limited. This happened for example in the Jewish Ashkenazi community, and among them in some orthodox groups, as we will further consider.

There are cases where religion did not play any role. For example, in Canada, the *Genetic Education Canada Knowledge Organization* (GEK-CO) was founded to increase awareness and genetic literacy among professionals and the population to prevent genetic diseases involving specific ethnic groups such as Ashkenazi Jews, French Canadians in the Quebec Region, persons of Cree ancestry, Aboriginal Manitoba and Newfoundland populations and families with Amish, Mennonite or Hutterite background (http://geneticseducation.ca/). The GEK-CO is a non-religious and non-profit organisation that was born out of the *Children's Hospital of Eastern*

Ontario, which is also a funding partner. By the way, they also promote the Canadian guidelines for maternity care and reproductive healthcare providers on pre- and post-conceptional reproductive carrier screening provided by *The Society of Obstetricians and Gynaecologists of Canada* (SOGC) (Wilson et al. 2016).

The *Jewish Genetic Disease* Consortium recommends that "all couples with ANY Jewish ancestry [that is, with Ashkenazi (German, French or Eastern European), Sephardic (Mediterranean), or Mizrahi (Persian/Iranian or Middle Eastern) ancestry], including interfaith couples, should have pre-conception carrier screening for all Jewish genetic diseases" (http://www.jewishgeneticdiseases.org/). This really means a number of genetic diseases. It is to note that, even if the advices concerning possible tests are not religiously embedded, the Consortium has launched educational programs to genetically educate Rabbis so that they may in turn inform young couples before marriage hoping they could follow an apt reproductive choice.

There are few data to evaluate the effectiveness of these programs or these recommendations. Nevertheless, the combination of the efforts made among Jewish congregations outcome the virtual elimination of certain genetic diseases, such as Tay-Sachs disease and Familial dysautonomia[5] among Jews of Ashkenazi ancestry (Lerner 2009).

The Religious Compulsion

Carrier testing for Tay-Sachs disease is considered the prototypic genetic population screening ethnically oriented. Tay-Sachs disease is an autosomal recessive disease caused by a mutation in the Hexosaminidase A (HEXA) gene, which results in the build up of the toxin GM2 ganglioside within cells. That mutation causes a progressive deterioration of nerve cells and of mental and physical abilities that begins around 6 months of age.

The screening was first launched among Ashkenazi Jews in northern America in the 1970s and then had a worldwide diffusion reaching thousands of Jews of Ashkenazi descent who were at high risk of giving birth to affected children, since the carrier rate among them highly exceeds that of other ethnic groups. Adhesion to the tests has been very high in this group and rapidly led to a 95% decrease in the disease incidence. Necessarily genetic screenings for recessive diseases involve the possibility of abortion or, in recent years, of preimplantation diagnosis. A state that would promote, or only favour, a genetic screening policy denying these possibilities, paradoxically, as it was in Italy until recently, would open serious contradictions.

The story of the carrier screening of this ethnically oriented disease is, however, strongly interlaced with religious suasion. Here the *Dor Yesharim* (it means "upright

[5]Familial dysautonomia is a genetic disorder that affects the development and survival of certain nerve cells. It is caused by mutations in the ELP1 gene, which provides instructions for making a protein that is found in a variety of cells throughout the body, including brain cells.

generation" and it comes from *Psalm* 112:2[6]) program enters the scene (http://doryeshorim.org). It was developed in 1983 "from one man's indescribable grief and heartache as he helplessly watched four beautiful children die of Tay-Sachs. Determined to ensure that no parent would ever endure the agony of losing a child to this devastating genetic disease, Rabbi Josef Ekstein galvanized a grass roots organization to eradicate Tay-Sachs forever". It is a screening program, which now is no longer limited to Tay-Sachs,[7] strongly pursued in the American Jewish Orthodox community, that has had a surprising success, as has been discussed in numerous papers (see the reference in http://doryeshorim.org/genetic-research-2/#research).

Adherents to this program undergo a premarital screening test. Carrier status is not revealed to any interested party and is known only to religious authority by which the marriage license depends. Weddings are usually combined and unions must be approved, but they are denied if both of the betrothed are carriers of a serious illness. This prevents the engagement according to a complex and peculiar procedure that involves the database of carriers, which is known only to the religious authority that decides. The *Dor Yesharim* program is an example of a community pretending to prevent the birth of a seriously ill person and not to enhance the procreative decision capacity of a couple or an individual. The community has the right/duty to establish that couples should not procreate given a risk to give birth to a severely sick child. Needless to say, this program has raised a number of ethical criticisms, ranging from strong paternalism to eugenics and stigma generator (see, for example, Rosen 2003; Raz and Vizner 2008). It is an example of a top-down approach characterised by the fact that even if the test was not compulsory, the possible positive result begets a chain of strongly paternalistic actions with an intense religious flavour. It is the religious governance of the community who spurs for genetic tests, who decides what to say and whether to say, and who "suggests" whether it is the case to get married.

Order Without Plan

A few years ago Castellani and colleagues reported that the incidence of Cystic fibrosis (CF)[8] had rapidly declined in a North-Eastern Italian region following the availability of the carrier test for mutations causing the disease (Castellani et al.

[6]"Their children will be mighty in the land; the generation of the upright will be blessed".

[7]It includes also Familial dysautonomia, Cystic fibrosis, Canavan disease, Glycogen storage disease, Franconi aenemia, Bloom syndrome, Niemann–Pick disease, Mucolipidosis type IV and Gaucher's disease. There is a Standard Ashkenazi Panel, a Sephardi/Mizrahi Panel, an Optional Ashkenazi Panel, and an Additional Screening.

[8]Cystic fibrosis is the most common life-limiting autosomal recessive disease among people of European heritage. It is caused by a mutation in the gene Cystic fibrosis transmembrane conductance regulator (CFTR). Such a mutation affects the chloride ion channel function resulting in

2009, 2016; Boniolo et al. 2012). Such accessibility was indirectly promoted by the University of Padua, although it was not supported either by local health authorities or by the Cystic Fibrosis Scientific Society or by the local Regional CF Referral Centre. Encounters were programmed with the gynaecologists working in the region and the test was offered by several laboratories at a nominal cost. However, over 200,000 persons have been tested, even if there was not the possibility to make a precise map of the reproductive choices of the individual couples who asked for the test. There were only data concerning the number of the newborn with CF and, as said, Castellani and colleagues observed a drastic decrease in the incidence of the disease. It is to be remarked that in Italy, at that age, while abortion was allowed under certain conditions (among which the foetus's severe pathological situations), preimplantation diagnosis was not legal.

As mentioned, this drastic decrease was not an effect of a top-down institutional policy but a sort of unintentional consequence at population level of the idea of the University of Padua to offer to the citizens the possibility to be tested at a nominal cost. And the individual citizens decided intentionally to act in this way. Of course this offer was done without any intention to modify the reproductive choices but only with the intention to empower the decisions of young couples. The adherence to the test has been surprisingly high and he individual reproductive choices led to the decrease of the incidence of CF: a sort of order (the unintentional collective decrease of newborn with CF) without a plan (no policies at all, non moral suasion by state or religion, no intention to obtain such an outcome), or, said differently, a collective result of intentional individual choices (the individual citizens who decided to be tested and the individual couples who decided the reproductive choice), of course due to the fact that the test was available at nominal cost.

A Deliberative Strategy for Population Carrier Screening: A Proposal

We begun with a question: "*Can population-based carrier screening be left to the citizens and in which way is this feasible?*". We have seen that there are policies totally top-down, be them instantiated by governments or by religious institutions, where the citizens' participation in choosing them and in choosing the diseases to be screened is virtually absent. We have mentioned that there could be a citizens' active role when a part of them is affected by a particular genetic disease and, therefore, such a part proactively try to implement, or support, or spur a policy. Finally we have also considered the case of Veneto-Italy, which to our knowledge is unique, characterised by a sort of latent policy which emerges as unknown, unexpected,

dysregulation of epithelial fluid transport in the lung, pancreas and other organs (see Castellani and Assael 2017).

and unintentional effect of individual independent choices, even if there has been an active role played by the University of Padova.

From all of this, it seems that the policies for carrier screening have been left to the citizens only if they were patients or in some way involved in the disease. But really this should or could be the only case in which citizens decide about which screenings and which policies? Only whether they are patients or patients' relatives? Really, as alternative, we should leave any choice to secular or religious institutions that decide more or less everything utilising a "William J. Clinton administration's formula for devising a national care policy [. . .]: *Gather the experts. Close the door. Design a policy. Roll it out. Reject criticism.*" (Ober 2008, p. 1; see Boniolo 2012).

On the other hand, one could say, those institutions "work for us". They themselves claim that they know better than us—the citizens—what to do and why. They choose the genetic disease to be screened; they provide the policy to be followed; they assess its instantiation also deciding its degree of compulsiveness; etc. Surely, for a citizen affirming his/her right of choosing about his/her health, this is a form of unacceptable paternalism, which could range from an extremely strong form, as in the case of *Dor Yesharim*, to a mildly strong form, as in the Italian or Cypriot thalassemia case. Is there another possible and more democratic avenue that could be pursued?

In the introduction, we have mentioned libertarian paternalism. As said, the main idea of this approach is to predispose the social scenario and the suitable nudges, thanks to choice architects, such that the citizen might be easily induced (rather, nudged) to act in the way that is thought best for him/her. But there is a problem: libertarian paternalism is still a form of institutional paternalism, although weak, rising the question of who (and why) is entitled to legitimize the institutional choices. If we were not able to give a plausible solution to this problem, we would be in the same situation as before: the institution (whichever it could be) tells us (more or less gently) what we should do and this, again, might not be accepted by individuals thinking that their will comes first. Yet, there is an alternative possibility, that some of us have already proposed (Boniolo et al. 2014; Boniolo and Rebba 2015): to introduce the deliberation. Let's see how.

An authoritative institution (whatever it could be and whatever could be its level) chooses statistically 5/6 sample groups, each of them composed of 20/30 individuals, out of the overall population. Each group has a sort of retreat in a given location for a bunch of days. During this period, the members of the groups receive a "sufficient" scientific information about genetic diseases, particularly those recessive, about the incidence of recessive genetic diseases in that country, and about possible reproductive choices. Here "sufficient"—intentionally a vague term—should be intended as the degree to which the group members acquire a clear, even if not professional, biological and clinical knowledge and understanding of the matters at issue. Moreover, also ethical notions should be given about the moral implications, both from an individual and collective point of view, of genetic tests and reproductive choices. Even some legal information should be provided about what is legal and illegal in that country. Finally religious information should be offered in order to furnish the right framework from a Christian, Islamic, Judaic, etc. point of view. All of this was complemented with basic information on how to correctly deliberate, that is, on critical thinking and on its use in deliberative contexts. After this training period, under the supervision of a facilitator the groups are asked to reach a conclusion on the necessity to implement a public preventive strategy and, in that case, which diseases should be concerned. A final group is then formed based on a statistical selection of some members and is offered the opportunity to retreat and to have a rehearsal of what already learnt about science, ethics, law, religion and critical thinking. Then, there should be again the deliberative discussion for the final position.

At this point the institution should accept what has been deliberated and act in a libertarian paternalistic way predisposing the right social design with the suitable nudges. In such a case, the libertarian paternalist approach would have been approved, decided and democratically legitimized by the deliberative process.

This structured process may be considered utopic, or at least unlikely to realise. Actually it is not. More or less an analogous deliberative process was implanted in UK to decide, as mentioned in the introduction, the moral plausibility of working with cybrids. That deliberative process was not without critics (Baylis 2009). These have pointed out some weaknesses concerning how the questions were posed to the chosen deliberating citizens and which kind of information was provided to them. Notwithstanding these flaws, the UK case deserves to be studied and to be used as a starting point to better plan a deliberative process concerning which genetic screenings should be made and how the institutional policy should be.

There is also another aspect to highlight. The UK deliberation was intended to know the UK citizens' point of view on the moral plausibility of the researches on cybrids. Actually, genetic diseases have some peculiarities. They have different incidences among ethnic groups, they may be prevalent in some areas, thus requiring different panels of mutations in different populations. In recent times it has also been suggested that new generation genome approaches might be cost-effective thus allowing population-based whole genome studies (Shivani et al. 2015), etc. This means that a unique and fixed screening panel for an entire population, especially if it is genetically heterogeneous, is rather unfeasible and would better require a screening panel adapted to the genetic homogeneity of the communities at issue. Moreover, we should underline that the deliberated libertarian paternalist policies could vary from offering only the possibility to be screened to an open suggestion concerning the available reproductive choices, etc. Needless to say, all of this should be taken into great account when the deliberative process is planned: an issue that, however, we skip here, being beyond our scopes.

Note, and this is the central point, that one thing is a top-down approach to health policies: the institution decides to be (strong or weak) paternalist and more or less mandatorily forces the citizens to adopt a certain health policy. A totally different thing is a bottom-up approach that legitimates the institution to adopt top-down initiatives: the citizens deliberatively decide to partially renounce to a small percentage of their autonomy and to give to the institution the permission to be weakly paternalist, since they accept that in certain cases the institution can do better, but they deliberate about what the institutions could be paternalist and in which way!

Thus, coming again to our question: "*Can population-based carrier screening be left to the citizens and in which way is this feasible?*". The answer could be: "*Yes, certainly it is. And it is feasible by means of a libertarian approach deliberatively chosen and legitimated by citizens after a well-formed decisional process. On the other way this is the only form of real democracy with real citizens' participation!*".

References

Abelson, J., et al. (2013). Public deliberation in health policy and bioethics: Mapping an emerging, interdisciplinary field. *Journal of Public Deliberation, 9*, 1–35.

Al-Gazali, L., Hamamy, H., & Al-Arrayad, S. (2006). Genetic disorders in the Arab world. *British Medical Journal, 333*, 831–834.

Angastiniotis, M. A., & Hadjiminas, M. G. (1981). Prevention of thalassaemia in Cyprus. *Lancet, 1*, 369–371.

Baylis, F. (2009). The HFEA public consultation process on hybrids and chimeras: Informed, effective, and meaningful? *Kennedy Institute of Ethics Journal, 19*, 41–62.

Beck, S. (1999). Blut, Schweiß und Gene – Anmerkungen zur Belebung der Künstlichkeit und zur Industrialisierung der Biologie. *Notizen zu Alltagskultur und Volkskunde, 14*, 11–16.

Bodenheimer, T. (1997). The Oregon health plan. Lessons for the nation. *New England Journal of Medicine, 337*, 651–655 and 720–723.

Boniolo, G. (2012). *The art of deliberating. Democracy, deliberation and the life sciences between history and theory*. Heidelberg: Springer.

Boniolo, G., & Di Fiore, P. P. (2010). Deliberative ethics in a biomedical institution. An example of integration between science and ethics. *Journal of Medical Ethics, 36*, 409–414.

Boniolo, G., & Rebba, V. (2015). Cancer, obesity and the legitimation of suggested lifestyles. A libertarian paternalism approach. *ecancermedicalscience, 9*, 588.

Boniolo, G., & Schiavone, G. (2015). Deliberation and democracy. In J. D. Wright (editor-in-chief), *International encyclopedia of the social & behavioral sciences* (pp. 61–67, 2nd ed.). Oxford: Elsevier.

Boniolo, G., et al. (2012). Cystic fibrosis carrier screening in Veneto (Italy) – An ethical analysis. *Medicine, Healthcare and Philosophy, 15*, 321–328.

Boniolo, G., et al. (2014). Libertarian paternalism and health care policy: A deliberative proposal. *Medicine, Healthcare and Philosophy, 17*, 103–113.

Boniolo, G., Schiavone, G., & Mameli, M. (2015). Moderate epistocracy for deliberative bioethics. *Cambridge Quarterly of Healthcare Ethics, 24*, 1–9.

Bozkurt, G. (2007). Results from the north Cyprus thalassemia prevention program. *Hemoglobin, 31*, 257–264.

Camporesi, S., & Boniolo, G. (2008). Fearing a non-existing Minotaur? The ethical challenges of research on cytoplasmic hybrid embryos. *Journal of Medical Ethics, 34*, 821–825.

Canali, S., & Corbellini, G. (2002). Lessons from anti-thalassemia campaigns in Italy, before prenatal diagnosis. *Medicina nei Secoli, 14*, 739–771.

Cao, A., & Kan, Y. W. (2013). The prevention of thalassemia. *Cold Spring Harbor Perspectives in Medicine, 3*, a011775.

Cao, A., et al. (1989). The prevention of thalassemia in Sardinia. *Clinical Genetics, 36*, 277–285.

Castellani, C., & Assael, B. M. (2017). Cystic fibrosis: A clinical view. *Cellular and Molecular Life Sciences, 74*, 129–140.

Castellani, C., et al. (2009). Association between carrier screening and incidence of cystic fibrosis. *Journal of the American Medical Association, 16*, 2573–2579.

Castellani, C., et al. (2016). Cystic fibrosis carrier screening effects on birth prevalence and newborn screening. *Genetics in Medicine, 118*(2), 145–151.

Haldeman-Englert, C. (2013). *Canavan disease*. MedlinePlus. http://www.nlm.nih.gov/medlineplus/ency/article/001586.htm

Hamamy, H. (2013). Community genetic services in Arab countries. *Middle East Journal of Medical Genetics, 2*, 6–10.

Infantino, L. (1998). *Individualism in modern thought: From Adam Smith to Hayek*. London: Routledge.

Kaback, M. M. (2001). Screening and prevention in Tay-Sachs disease: Origins, update, and impact. *Advances in Genetics, 44*, 253–265.

Khorasani, G., et al. (2008). Results of the national program for prevention of beta-thalassemia major in the Iranian Province of Mazandaran. *Hemoglobin, 32*, 263–271.

Lerner, B. H. (2009). When diseases disappear—the case of familial dysautonomia. *The New England Journal of Medicine, 361*, 1622–1624.

Matalon, R., & Michals-Matalon, K. (1993–2017). Canavan disease. 1999. In R. A. Pagon, et al. (Eds.), *GeneReviews*. Seattle, WA: University of Washington. Available from: https://www.ncbi.nlm.nih.gov/books/NBK1234

Mosconi, P., et al. (2015). Cystic fibrosis: To screen or not to screen? Involving a Citizens' jury in decisions on screening carrier. *Health Expectations, 18*, 1956–1967.

Mosconi, P., et al. (2016). Steering Committee; Scientific Committee. Involving a citizens' jury in decisions on individual screening for prostate cancer. *PLoS One, 11*, e0143176.

Ober, J. (2008). *Democracy and knowledge. Innovation and learning in classical Athens*. Princeton, NJ: Princeton University Press.

Raz, A. E. (2009). Can population-based carrier screening be left to the community? *Journal of Genetic Counseling, 18*, 114–118.

Raz, A. E., & Vizner, Y. (2008). Carrier matching and collective socialization in community genetics: Dor Yeshorim and the reinforcement of stigma. *Social Science & Medicine, 67*, 1361–1369.

Rosen, C. (2003). Eugenics—Sacred and profane. *The New Atlantis, 2*, 79–89.

Samavat, A., & Modell, B. (2004). Iranian national thalassaemia screening programme. *British Medical Journal, 329*, 1134–1137.

Shivani, B., et al. (2015). Changing trends in carrier screening for genetic disease in the United States. *Prenatal Diagnosis, 35*, 931–935.

Stafler, P., et al. (2016). The impact of a national population carrier screening program on cystic fibrosis birth rate and age at diagnosis: Implications for newborn screening. *Journal of Cystic Fibrosis, 15*, 460–466.

Sunstein, C., & Thaler, R. (2003). Libertarian paternalism is not an oxymoron. *University of Chicago Law Review, 70*, 1159–1162.

Taddeo, S., & Robert, J. S. (2014). *Hybrids and chimeras: A consultation on the ethical and social implications of creating human/animal embryos in research (2007), by the HFEA*. https://embryo.asu.edu/pages/hybrids-and-chimeras-consultation-ethical-and-social-implications-creating-humananimal-embryos

Tadmouri, G. O., et al. (2009). Consanguinity and reproductive health among Arabs. *Reproductive Health, 6*, 17.

Thaler, R., & Sunstein, C. (1993). *The winner's curse: Paradoxes and anomalies of economic life*. Princeton, NJ: Princeton University Press.

Thaler, R., & Sunstein, C. (2003). Libertarian paternalism. *The American Economic Review, 93*, 175–179.

Thaler, R., & Sunstein, C. (2008). *Nudge: Improving decisions about health, wealth, and happiness*. New Haven, CT: Yale University Press.

Wilson, R. D., et al. (2016). Joint SOGC-CCMG opinion for reproductive genetic carrier screening: An update for all Canadian providers of maternity and reproductive healthcare in the era of direct-to-consumer testing. *Journal of Obstetrics and Gynaecology Canada, 38*, 742–762.e3.

Public Health and Inequities Under Capitalism: Systemic Effects and Human Rights

Joan Benach, Juan Manuel Pericàs, Eliana Martínez-Herrera, and Mireia Bolíbar

All authors are members of Health Inequalities Research Group, Employment Conditions Knowledge Network (GREDS-EMCONET) and the JHU-UPF Public Policy Center at the Department of Political and Social Sciences, Universitat Pompeu Fabra in Barcelona. Joan Benach and Juan Manuel Pericàs are also members of the Transdisciplinary Research Group on Socioecological Transitions (GinTRANS2), Universidad Autónoma in Madrid. Juan Manuel Pericàs is also a researcher at the Translational Research Group on Infectious Diseases of Lleida (TRIDLE), IRB Lleida, Hospital Universitari Arnau de Vilanova, Lleida, Spain. Eliana Martínez-Herrera is also a researcher at the University of Antioquia in Medellín, Colombia.

J. Benach (✉) · J. M. Pericàs
Health Inequalities Research Group, Employment Conditions Knowledge Network (GREDS-EMCONET), Universitat Pompeu Fabra in Barcelona, Barcelona, Spain

JHU-UPF Public Policy Center at the Department of Political and Social Sciences, Universitat Pompeu Fabra in Barcelona, Barcelona, Spain

Transdisciplinary Research Group on Socioecological Transitions (GinTRANS2), Universidad Autónoma in Madrid, Madrid, Spain
e-mail: joan.benach@upf.edu; jmpericas.lleida.ics@gencat.cat

E. Martínez-Herrera
Health Inequalities Research Group, Employment Conditions Knowledge Network (GREDS-EMCONET), Universitat Pompeu Fabra in Barcelona, Barcelona, Spain

JHU-UPF Public Policy Center at the Department of Political and Social Sciences, Universitat Pompeu Fabra in Barcelona, Barcelona, Spain

Epidemiology Research Group, National School of Public Health, University of Antioquia (UdeA), Medellín, Colombia
e-mail: eliana.martinez@udea.edu.co

M. Bolíbar
Health Inequalities Research Group, Employment Conditions Knowledge Network (GREDS-EMCONET), Universitat Pompeu Fabra in Barcelona, Barcelona, Spain

JHU-UPF Public Policy Center at the Department of Political and Social Sciences, Universitat Pompeu Fabra in Barcelona, Barcelona, Spain
e-mail: mireia.bolibar@upf.edu

J. Vallverdú et al. (eds.), *Philosophical and Methodological Debates in Public Health*, https://doi.org/10.1007/978-3-030-28626-2_12

Abstract The chapter aims to reflect on the global systemic effects of capitalism on health and health inequities posing two questions: Is it possible to understand capitalism's systemic effects as a whole? Can we know the multiplicity of all these interrelated impacts on eco-humanity and on health?

The text claims that capitalism, as an economic and power system, impregnates the social organization of all human domains, thus permeating humanity and the many facets of the health-disease production process. Then, it presents the example of the agrofood, showing how the capitalist agro-industry has set an extractive, predatory, commodity-driven and inequity-generating system of food production and distribution that gives autocratic power to agrofood oligopolies competing to lead a global capitalist market and ultimately produces both health-damaging hunger and obesity.

It uses this example to further reflect on the extension of commodification of human life under neoliberal capitalism and the complex set of processes and mechanisms by which it directly or indirectly affects our minds and bodies. In this sense, the chapter highlights the great importance that social and environmental media play in the production of health, to the extent that even behavior and the micro psychobiological effects on specific organs, cells and neurons should be understood under the light of history, power relations, economic systems, culture and ideology.

Finally, the chapter translates such reflections into a criticism of how current global political institutions frame the human right to health. It claims that in order to resist the systemic pressures of capitalism, to democratise states and international institutions and to control and regulate the large oligopolies, it is essential to rethink the hegemonic discourse of the right to health so that it becomes a first-class human right intrinsically coupled to the other social, political and economic rights.

Keywords Capitalism · Systemic effects · Social production of health and disease · Health equity · Human right to health

What Is Capitalism and What Are Its Systemic Effects?

Capitalism can be understood as a continuously mutating historic "organism", an economic and power system, dynamic and expansive, which is based on the constant dispossession of the popular masses, the destructive colonization of the planet, and the appearance of periodical crises that only it that it can only move geographically and resolve momentarily through domination, violence, the concentration of the means of production and the proletarianization of humanity. Capitalism is a system that demands continuous economic growth; the endless accumulation and circulation of capital based on expropriation of the majority of the population's means of subsistence; while being at the same time a power system that is exercised through the use of violence and population control. Capitalism has not only transformed human history and the planet itself, but also has colonized the integrity of social and cultural life, as well as consumption and leisure patterns. Its evolution in the twenty-first century will decide a future in which encompassing a high risk of collapse and extinction.

In the last decades, the triumph of neo-liberal capitalism has been extraordinarily broad and profound. Its victory has not only been economic and political, but it has broken class consciousness, destroying social links and fragmenting identities. Mercantilization has extended from the microcosms to the macrocosms in almost every realm and in all things: healthcare, education, nature, knowledge, culture, art, the body and human relationships (Wallerstein 1983; Varoufakis 2017). Genes, bacteria, seeds, tissues and genetically modified animals are patented; organs are bought and trafficked; uteruses, families and even friends and lovers are rented; and land lots are sold on the Moon and on other planets.[1,2] Capitalism pretends to control every source of workers' labor force and to have a complete domination over everyone's minds and bodies, specially women's, on whose shoulders has fallen, in great measure, the reproduction of life: child-bearing and child-raising, affective links, cleaning chores and care activities (Federici 2014). In short, the capitalist system and its diverse contradictions produce multiple systemic manifestations on ecology, production, transportation, the market, social relationships, daily life, consumption, education and other factors that—as we defend in this writing—are essential to be understood if we are to grasp the extent and distribution of eco-humanity's health and disease.

Is it possible to understand capitalism's systemic effects as a whole? Sociologist E. O. Wright has synthesized his critique of capitalism in eleven points, from exploitation, lack of autonomy and freedom, systematic and unnecessary vulnerability and suffering that hinder human fullness, commodification of shared values, generation of efficiencies, lack of opportunities and consumerism; to the promotion of militarism and imperialism, ecological destruction and the corrosion of communities and democracy (Wright 2010). Can we know the multiplicity of all these

[1]There are three types of goods in the world: universal, general and collective. Universal goods are represented by unique things such as the Sun, mountains or Beethoven's 9th Symphony from which everyone can benefit; general goods are those such as food, housing, sanitation, education, etc. which we have to generalize in order for humanity to be complete; collective goods are things like an automobile which we must share and rationalize, those whose benefits everyone should enjoy equally but can't be generalized without endangering the existence of general and universal goods. See: Santiago Alba Rico. Que haya ricos, ¿no es un derecho de los pobres? Rebelión. 29-10-2009. Accessible at: http://www.rebelion.org/noticia.php?id=94115

[2]Throughout history, social classes have appropriated general and collective goods, but capitalism has been increasingly trying to appropriate universal goods as well. In 1895, Cecil Rhodes, an English imperialist, businessman and founder of De Beers company (who owns 60% of world diamonds), looked with anger at the stars through his window, "so clear and so distant", so far away from that imperial appetite "that wanted but couldn't annex them". North-American entrepreneur Dennis Hope registered the Moon in his name in 1980. Hope took advantage of a legal loophole although there is an international treaty that indicates that no country can make a property claim over the Moon or another celestial body, this doesn't explicitly forbid individual or company claims. The satellite has been divided into lots to be sold through his company Lunar Embassy. The same could happen with Mars, Mercury and Pluto. Accessible at: http://www.lunarembassy.com/about/

interrelated impacts on eco-humanity and on health? The answer should be no for at least three reasons. *First*, because of the huge complexity of the determinants and causal processes involved in the process of health-disease production, which represent enormous research challenges. Thus, a big part of structural determinants in the political and socio-economic spheres are "invisible causes", difficult to comprehend and analyze, and entangled in a complex network of multilevel causation (Vallverdú 2016). Furthermore, the bigger and further away that the "casual network" is located, the more probable it is that historically contingent contextual effects will make appearance (Krieger and Smith 2016). *Second*, because we need more and better socio-sanitary data and indicators on key processes and problems that allow us to determine the relationship between capitalism and health and, as of today, these are limited, inaccessible or inexistent. And *third*, due to the urgent need to count with critical scientists that carry out systemic analysis—a sort of "total science"—in transdisciplinary research groups, alternative to the progressively reductionist and mercantile current system of hegemonic research that simplifies reality and privatizes knowledge, turning it into another part of a classist system often conceived as a means to justify the existing social order. Due to this, we require new or renewed epistemic approaches—which are inevitably ethic and political in nature as well—that overcome the ideology of purported scientific neutrality (Farrés 2016), and that recognize health as a product of social, political and economic organizational systems.

Impacts of Capitalism on Public Health

Capitalism is the ultimate cause of the current global pathogenesis that enters our bodies and minds almost without us noticing. Pharmaceutical companies invent diseases and syndromes while they identify treatments for all sorts of maladies, addictions, neurosis, disorders, worries, pains, humiliations and fears caused by capitalism itself. The body is analyzed, fragmented, commercialized and finally sold as a commodity. Health is for sale, the bodyminds are an ideal commodity to be defined, classified, monitored, repaired, controlled, and finally sold and bought. There seem to be no limits at all. "To buy health" in every field of life: childhood, adolescence, sexuality, work, food, the cult of the body, sports, leisure, ageing and death. Health and body thus appear to be the last frontiers of voracious and relentless capitalism.

Not having an exhaustive—or at least sufficiently complete—systemic knowledge of how capitalism permeates humanity and damages health does not mean we get to ignore all that we do know. Nowadays we have quite extensive knowledge over processes and mechanisms that directly or indirectly affect our bodyminds in many diverse areas and due to wide-ranging causes. Just to mention some of them: from the genocidal death and destruction caused by hundreds of imperialist wars, to colonialism's and neocolonialism's diverse and multi-faceted damages; from exploitation and labor precarization, to multiple harms generated by toxins and labor risks, to all sorts of domination and precarization due to class, gender, ethnicity, national identity, cultural or sexual differences, all which cause great inequities in health; from the privatization, commodification and medicalization of socio-sanitary care to the (iatrogenic) harm generated by the technological-pharmaceutical-chemical industry or the penury of basic medical care in poor countries.

Economic Oligopolies: The Example of Agro-Foods

Big corporations are economic entities that possess great wealth, to the point that 69 of the top 100 economic entities in the world are big corporations. It is estimated that the first 10 hold more wealth than the 180 least affluent countries combined.[3] These corporations are grouped in a number of conglomerates concentrated around the food sector, energy products, raw materials and services.[4] Its social, political and economic power is enormous (Noam Chomsky has called them "Private Tyrannies"): they hire or outsource millions of job posts, induce people to buy a great deal of the goods and services that we consume, exert an enormous pressure over governments and institutions, and increasingly speculate in a complex and opaque financial economy.[5,6] Following capitalism's endemic process of industrial oligopolization, the world's food in the hands of just a few agro-food corporations (the "Big Food"), whose basic objective is not to produce food but to sell the biggest possible volume of commodities and thus obtain the largest possible benefit.[7]

---

[3]See the democratic organization after global justice. Accessible at: http://www.globaljustice.org.uk/sites/default/files/files/resources/corporations_vs_governments_final.pdf

[4]Seven companies conform "Big Oil"; five companies—which could soon be less—"Big Tobacco"; thirteen companies are part of "Big Pharma"; four are part of what's called the "Big Four" (consulting and auditing); six big agrochemical firms conform the "Big 6"K six are also part of the "Big Media"; ten, "Big Food"; three, "Big Soda"; four, "Big Chocolate", etc.

[5]Though nobody knows its true size, it is estimated that the size of speculative financial economy could be 125 times larger than the size of the real productive economy. See: G. López. *La economía especulativa supera en más de 125 veces el dinero en metálico.* May 10, 2016. Accessible at: www.elsalmoncontracorriente.es

[6]It is estimated that there are US$7.6 trillion of individual financial assets hidden in fiscal paradises (Jean Ziegler called them "pirate paradises") worldwide. Up to 30% of the entirety of African financial assets is in fiscal paradises, at an annual loss of US$14 billion. That amount would allow guaranteeing the education, sanitation and lifesaving of four million African children per year. See: Oxfam. *An economy for the 1%.* January 18th, 2016. Accessible at: https://d1tn3vj7xz9fdh.cloudfront.net/s3fs-public/file_attachments/bp210-economy-one-percent-tax-havens-180116-en_0.pdf

[7]Even though hundreds of products and brands fill supermarket shelves, all of them belong to a handful of conglomerates: Nestlé owns more than 60 food brands (Nestea, Nesquik, Kit Kat, etc.), cosmetics (L'Oréal), or textiles (Diesel, Giorgio Armani, Ralph Lauren, etc.). Then we have Procter and Gamble, Unilever, Kraft Food, Kellogg's, Mars, PepsiCo, Coca-Cola, Johnson & Johnson and General Mills. The biggest ten generate income of more than US$1.1 billion everyday; they employ millions and represent approximately 10% of the world's economy.

The global food industrial complex is articulated through a number of restaurant chains, supermarkets, pre-cooked meals manufacturers; the purchase, transportation and selling of cattle, fish and vegetable products; bio-genetics, seed production, insecticides, herbicides, and fertilizers (de Sebastián 2009a). The market of seeds and chemical products is currently dominated by six agro-food companies, but in the near future these could be further reduced to three mega-corporations (Bayer buying Monsanto, Dow merging with Dupont, and ChemChina buying Syngente) which would control almost 60% of all seeds, close to 70% of agrochemical pesticides and nearly all transgenic patents.[8] Agroindustry's corporate consolidation could further increase with its investments in *big data*, robotics and technology and the control of patents and intellectual property, resulting in an increasingly harmful and vulnerable "single farm model", which would exclude small farmers who total 90% worldwide and provide more than 80% of poor countries' food.

Capitalist agro-industry is an extractive, predatory, commodity-driven and inequity-generating system, incompatible with social justice and whose negative impacts over health and ecology are increasing (Riechmann 2003). This agro-food model could, according to FAO, produce enough food to feed 12 billion people but instead produces malnutrition and obesity at the same time. It is a productivist agriculture, based on machines and technologies that consume oil, fertilizers and pesticides intensively, based on monocrops and with a tremendous power over prices, availability of seeds and the entire model of production, distribution and consumption. The food oligopolies determine how much is produced, what is to be eaten, where, how and at what price. We can thus summarize its main power mechanisms: First, it plunders foodstuff from hungry places in the Global South to others of great abundance in the Global North (Schuldt 2004). Second, it fixes prices and controls food trade along the production-consumer chain: from seeds and pesticides, to storage facilities, transportation, etc. (Garcia 2016) And third, it speculates with the value of food, which means that a great deal of staple crops are used as bets in stock exchanges (Jean Ziegler, former UN Special Rapporteur for the Right to Food called this "banking banditry"), to the point that financial trading of wheat related stocks move 50 times more money than its production. This even creates food crises on a world scale, particularly affecting such countries which, following neo-liberal policies in the context of Structural Adjustment Plans, have put aside their national agrarian policies, abandoning local production. FAO recommends that investments in the agrarian sector total at least 20% of the national budget, but the IMF and the World Bank propose a number closer to 4%.

Hunger affects around 800 million people (at least 1 out of 9 worldwide), most of them in the Asia-Pacific Region (511) and in Africa (232), the continent with the

[8]J. Vidal. Farming mega-mergers threaten food security, say campaigners. The Guardian. 26/09/2016. Accesible en: https://www.theguardian.com/global-development/2016/sep/26/farming-mega-mergers-threaten-food-security-say-campaigners; J. Vidal. A switch to ecological farming will benefit health and environment—report. The Guardian. 02-06-2016. Accessible at: https://www.theguardian.com/environment/2016/jun/02/a-switch-to-ecological-farming-will-benefit-health-and-environment-report

biggest percentage of hungry people (23% of the population) and with the biggest number of vulnerable countries (Central African Republic, Chad, Zambia, Sierra Leone and Madagascar). Hunger is a permanent earthquake, a crime against humanity. But hunger is not a problem of production or supply but one of poverty, power and inequality in resources and absence of democracy. Hunger has always been present throughout history but today it's entirely avoidable. A child that dies of hunger in the twenty-first century is a murdered child.[9] Just as there's explicit hunger there's also "hidden hunger", the type that affect people who satiate their feeding needs with rice, corn or wheat but do not obtain enough nutrition, thus making essential the availability of vitamins, oligo-elements and other micronutrients. The worst effect of food insecurity is not in fact deaths from inanition, but malnutrition affecting a third of poor countries' population and the co-adjuvant diseases that it causes. Lack of food and malnutrition do not leave permanent sequelae in adults, but in infants it produces development issues: the immunological system is weakened, and growth alterations and deficit cognitive processes are generated.

The other side of the capitalist coin is overfeeding, which has become a global public health problem as important as hunger. During the last four decades obesity has multiplied: the number of obese people has grown from 105 million in 1975 to 641 million in 2014, and the figure reaches 2 billion adults when overweight is also accounted (41 million among children under 5 years) (WHO 2016).[10] There are more obese people (13% of the world's population) than underweight (9%) and everything indicates that the situation will continue to worsen. The majority of the world's population lives in countries where obesity and overweight take more lives than the pondered insufficiency. Overweight and obesity are the fifth largest death risk in the world, killing more than three million adults annually. Forty-four percent of the burden of diabetes, 23% of the burden of ischemic heart disease and between 7 and 41% of the burden of some cancers can be attributed to obesity and overweight. Additionally, obesity and overweight are closely related to geographical social inequities in cardiovascular risk factors such as high blood pressure, cholesterol or insulin resistance.[11]

The powerful food industry spends huge amounts of resources to generate obesogenic contexts, stimulating the consumption of processed foods, fast food, sugary drinks and a whole range of "normal" foods to make them more enticing

[9]J. Ziegler. Destrucción masiva: Geopolítica del hambre. Barcelona: Península, 2012:19,52.

[10]If current trends persist, it is estimated that in 2025 the global prevalence of obesity will be 18% in men and 21% in women. A study by a Lancet Commission analyzed data from 1698 studies based on a population of 19.2 million men and women under 18 years of age in 186 countries, comprising 99% of the world's population. See: NCD Risk Factor Collaboration (NCD-RisC). Trends in adult body-mass index in 200 countries from 1975 to 2014: a pooled analysis of 1698 population-based measurement studies with 19.2 million participants. Lancet 2016; 387: 1377–96.

[11]Obesity almost always appears in the form of social inequities according to social classes, race/ethnicity and geographic distribution.

(De Vogli et al. 2014).[12] Besides a sophisticated grid of marketing and social relations, the three pillars of obesity's "Holy Grail"—over which the agro-food industry is sustained—are salt, fats and sugar. Today, research and technology have allowed us to know the exact amount of each ingredient to create appetizing and addictive foods that "hook" the consumers and incite them to continue eating (Moss 2003). As explained by journalist Michael Moss: "As a result of this $1 trillion-a-year industry, *one*-in-*three adults, and one*-in-*five children*, are now *clinically obese*." (Moss 2003).

Both hunger and obesity have political causes and therefore should be solved through politics. They are not, therefore, a natural and automatic result of technological evolution, the market, the prices, etc., but instead are sustained by the food production and distribution systems' capitalist basis, which gives autocratic power to agro-food oligopolies competing to lead a global capitalist market. Their purpose is not to produce food, its objective is to increase its benefits through a complete dominion over the integral agro-food cycle, allowing them to speculate with the price of food and increase its sales (de Sebastián 2009b; Nestle 2002).[13]

"Capitalist Cells", or How the System Enters into Our Bodies

One of the most cited causes to explain disease and justify differences in health is the so-called "biological determinism" (Lewontin 1984). It is an ideology, constantly spread through mass media and by many respected scientists, according to which genetic and biological agents would be the main "culprits" of disease as well as many other life and health situations. It is not by chance that in recent years we've witnessed a profuse and erroneous dissemination of the highly repeated expression "it's in the DNA" to refer to different things that seem to be highly stable or unchangeable. Although there's hardly a doubt that genetic factors are important and deserve to be considered within issues of collective health, in truth they play only a relatively minor role in the social production of disease and in the genesis of health inequities and community health as a whole.

[12]For example, it is estimated that Coca Cola's yearly advertising budget is bigger than the entire budget of WHO. See: Chopra et al. A global response to the global problem: the epidemic of over nutrition. Accessible at: www.who.int/bulletin. 2002.

[13]The capitalist agro-food model generates other negative consequences that could be briefly put as follows: (a) lack of land, its hoarding, and loss of biodiversity; (b) massive precarization of the working conditions of poor farmers subject to labour exploitation; and (c) the toxicity of chemical residues in foodstuff. These chemicals are favoured by the manipulations, pressures and mafia tactics of big chemical corporations. The alternative to that model is to fight for the food sovereignty of the people, and for a social, sustainable and democratic model of social agriculture. See: J. Riechmann. Cuidar la T(t)ierra. Políticas agrarias y alimentarias sostenibles para entrar en el siglo XXI. Barcelona, Icaria, 2003.

Several reasons sustain this asseveration. *First*, diseases whose origin is exclusively genetic, such as muscular dystrophy or Huntington's disease, only represent a small proportion of society's public health. "It is a fallacy of biological determinism to say that if differences are in the genes, no change can occur" (Lewontin 1991). *Second*, biological factors do not act in an isolated manner but are constantly interacting with the environment: an initial biological or genetic disadvantage can or cannot be compensated through an adequate change in the social medium. In fact, despite numerous researchers pointing out for decades that genetic factors exert a notable influence over the human conduct, there's been a failure in obtaining the necessary evidence that shows how specific genes cause differences in intelligence, personality, socially reproachable conducts or psychiatric disorders (Joseph 2015). *Third*, genetic predisposition almost always produces avoidable effects. For instance, causes of death or risk factors as important as coronary disease, lung cancer, hypertension or obesity, change their distribution among communities and social classes along time due to mainly historical and social causes. *Fourth*, there are health inequities among different social classes for many different diseases, which makes us claim the great importance that social and environmental media play in the production of health. And *finally*, our current knowledge of genetics and its relationship with disease is still very much incipient. As a matter of fact, the interaction with the environment is so complex that we'll probably never be able to predict or cure many diseases through genetic techniques alone.[14] Each human being is born, lives, works, interacts with others, sickens and dies influenced by the social milieu that surrounds them. We are social animals, not biological machines isolated from the rest of society. That does not mean that biology or genetics are unimportant, but only that they are not the most important public health issue, and that they shouldn't be approached as isolated from their environment.

In truth, health-related processes are the result of a wide group of interrelated causes that, in practice, are very hard to separate (Lewontin and Levins 2007). For example, when an immigrant woman arrives to a hospital's emergency room due to chest pain underlying a heart attack, it is her body expressing all the problems and risk factors accumulated throughout her life. She, as a human being, reflects in her biology and psychology her own personal history and the history of her social class, her gender and social collective, her community and even the country to whence she originally came from.

Marx and Engels had already pointed out that means of production not only referred to the physical existence of individuals, but also to a certain way of expressing people's existence, to a "mode of life" (Marx and Engels 1978). Likewise, sociologist Pierre Bourdieu employed the concept of *habitus* to express that "history turns into nature", turns into bodies (Bourdieu 1977). And social epidemiologist

[14]Kenneth M. Weiss. "Genetists are making false promises to the public. We cannot expect to cure every disease knowing only genes (...) The great amount of implicated genes and the diversity of their interactions make it impossible both to predict that a person will suffer a given disease as to intervene on genes to stop this". Interview at El País, March 27 2001.

Nancy Krieger has rightly pointed out that "people literally embody and biologically express experiences of economic and social inequality, from in utero to death". Social inequities are thus a result of a wide spectrum of health-related issues that get under our skin (Krieger 2001). We can definitely say that our minds and bodies express eco-sociology, *nothing in biology makes sense except in the light of history* (Krieger 2013).

In order to illustrate this point, let's take a look at a single example related to prenatal environmental effects. During the Dutch Hunger Winter of 1944, the Nazis diverted all foodstuffs from the Netherlands to Germany, so the Dutch population in general and pregnant women in particular were suddenly left close to inanition. A wide variety of studies have shown how fetuses that were in the womb during that time developed a calorie-saving metabolism due to the brutal nutritional deficiency conditions. As a result, many decades later these people have been more prone to develop metabolic diseases such as diabetes and obesity, as well as high blood pressure, etc. (de Rooij 2006) Those people's bodies "remember" the history they suffered while in their mothers' wombs.

This socio-historic approach to the causes of health and disease allows us to understand, as was stated by the great ecologist and social activist Richard Levins, that in any given moment we can speak of the harm done to interconnected bodyminds that suffer psychobiological changes related to history, ecosociology, power relations, the economic systems, culture and ideology, producers of capitalist organs, cells and neurons.[15]

Similarly, conditions under which the working force sells itself in a capitalist labor market define the health of the working population. Working conditions act over an individual's glucose cycle since the patterns of effort and rest depend more on economic decisions made by businessmen than on the workers' self-perception of their metabolic flow (Lewontin and Levins 2007). Likewise, the experience of unemployment generates anxiety and alcoholism, and this in turn produces a greater risk of cirrhosis and biological harm over the body of an unemployed person. A public health wide view that allows the hepatic cell, the hepatocyte, or the cellular enzymes that reflect the liver diseases, to link with the cyclic crises inherent to capitalism; capitalism's tendency to produce unemployment and unemployment benefitting businesses by creating a labor force reserve that encourages competition among the working population and allowing the degradation of working and employment conditions for the benefit of capital. Under that vision, it wouldn't be exaggerated to propose the existence of a "capitalist liver" or a "proletarian lung".

[15]"Racism becomes an environmental factor affecting adrenals and other organs in ways that tigers or venomous snakes did in earlier historical epochs. The conditions under which labor power is sold in a capitalist labor market act on the individual's glucose cycle as the pattern of exertion and rest depends more on the employer's economic decisions that on the worker's self-perception of metabolic flux. Human ecology is not the relation of our species in general with the rest of the nature, but rather the relations of different societies, and the classes, genders, ages, grades, and ethnicities maintained by those social structures. This, it is not too far-fetched to speak of the pancreas under capitalism or the proletarian lung". See: Lewontin and Levins (2007, p. 37).

Health as a Human Right and Systemic Approaches

Never before in history have so many people demanded and defended human rights. At the same time, however, under the current capitalist system, the right to health and interconnected social rights are not guaranteed or adequately upheld. Why? How is this possible?

The capitalism of the twenty-first century is not only a mode of production but also a cultural regime and a mode of civilization that as seen before embeds itself in our bodies and minds, in our behaviors, in our moral perspective and in the ideology with which we define and interpret our rights. Indeed, the history of human rights is the history of the victors. Human rights are only invoked when they serve the logic of power of those invoking them. As the Portuguese sociologist Boaventura de Sousa Santos notes: "When, starting in the middle of the nineteenth century, human rights discourse separated from the revolutionary tradition, it began to be understood as a depolitized grammar of social transformation, a kind of antipolitics" (de Sousa Santos 2014). This hegemonic view of human rights limits the ways in which we must understand them.

Faced with the idea of human rights as an emanation of the supposed ideal of European humanity, varying critiques have shown that this conceptualization of human rights is neither appropriate nor democratic in the real world. Also Sousa Santos says: "the Universal Declaration of 1948, [...] was made without the participation of most people in the world; in the exclusive recognition of individual rights, with the only exception being the collective right to self-determination that, moreover, was restricted for peoples subjected to European colonialism; in the priority given to civil and political rights over economic, social and cultural rights, and in the recognition of the right to property as the first and, for many years, the only economic right" (de Sousa Santos 2002).[16]

Under twentieth century capitalism, human rights were distorted in the name of "progress," "development," "growth," and "modernization", putting forward hidden conservative interpretations of human rights associated with liberal democratic states and neocolonial, neoliberal, neo-imperial, neo-patriarchal, and neo-extractivist policies. But rights that truly serve humanity must not only be credible and applied in practice. Human rights must also avoid its imposition under the perspectives of the most powerful social classes, dominant States, corporations, by patriarchy, whiteness, heteronormativity, national dominant views, and other forms of oppression and hegemony (Monedero 2015). There is a vital need for critical alternatives. For example, the anthropologist, engineer, and ecologist Yayo Herrero has reinterpreted the first article of the Declaration of Human Rights according to an ecofeminist

[16]The Universal Declaration of Human Rights of December 10th 1948 was nothing else than a reformulation of the Declaration of the Rights of Man and Citizen of 1793. The French revolution was a radical, social and republican democratic revolution, that was crushed by the bourgeois reaction and eliminated during the nineteenth and a big part of the twentieth centuries. See: Domènech (2009).

perspective: "All human beings are born vulnerable and defenseless in the bosom of a mother, and we will become free and equal in dignity and rights as long as we receive a huge amount of attention, care and affection, which must be provided by men and women from other generations in a civilizing task without which our species cannot exist. If we receive them, we can become endowed with conscience and reason that allows us to live fraternally with one another, aware that we inhabit a planet that has physical limits, that we share with the rest of the living world, and that we are obliged to keep (Herrero 2017)."

In the same vein, the right to health must also be reimagined beyond the current hegemonic paradigm through alternative and interdependent views (Mann et al. 1994). What do we mean when we refer to health as a "universal right"? Is health just another human right? As we have seen, the right to health has been recognized in multiple treaties from the United Nations Universal Declaration of Human Rights ratified in 1948, the International Treaty of Economic, Social, and Cultural Rights (ratified in 1966 and implemented in 1976) and other agreements that make up the International Charter of Human Rights. Viewing the right to health through this international legal perspective, we can distinguish four relevant characteristics. *First*, the historical construction of rights. The right to health was conceived in mid twentieth century postwar Europe after much struggle and negotiation, with the British National Health Service (NHS) serving as the pioneering example of a national health system with universal coverage. *Second*, the inability to generalize the right to health. There have been many critiques of the historical and theoretical framework of human rights and the extension of these rights to other regions of the world. *Third*, the reduction of health to healthcare. The right to health cannot be limited to the right to health care.[17] The right to health depends upon other fundamental and intertwined social rights (the social determinants of health) that, through both active and passive concessions, in the last few decades have been restricted, weakened or even eliminated altogether in the expansion and entrenchment of neoliberal capitalism. *Lastly*, the value of health as a human right. Economic, social and cultural rights are considered second generation human rights, which implies being considered second-class human rights and being pushed into the background (Vasak 1982). Its ratification and compliance are in the hands of each State, yet there is no jurisprudence that compels countries to comply with a comprehensive vision of health that includes all collective health and equity in health.

These considerations make demonstrate certain imperative assumptions of today's discourse on health as a human right, and to understand why the right to

[17]Considering that the responsibility to realize the right to health falls upon "the healthcare system", having more pharmaceuticals and more healthcare centres denotes a restricted conception of health as the accessibility of services centered around healthcare professionals and managers. We agree with Paula Braveman, who insists in calling this type of services "medical care" instead of "health care". See: Braveman P. Social conditions, health equity, and human rights. Health Hum Rights. 2010;12(2):31–48.

health is often not so well analyzed. Three are the main questions to be considered. *First*, how can health be a right since there is no binding legislation demanding this right? *Second*, how can the state of being in good health be a right, when there is no way of ensuring that everyone does have good health? And *third*, why do we need to think of health rather than health care as a right, since only the latter is under the direct control of policy making? The Indian economist and philosopher Amartya Sen has provided some answers to these specific questions (Sen 2008). For him the idea of right does not have to be necessarily thought as a legal issue but rather under the tradition of rights in terms of social ethics, i.e., "what a good society must have." The feasibility question is based on a common confusion about what can or cannot be a right.[18] The acceptance of health as a right for all is a demand to take action to promote that goal, going beyond the so-called "first-generation rights". The policy question points to the fact that good health depends on health care. Yet, as mentioned, good health goes beyond legislating good health care, as it mainly depends on the social determinants of health. Thus, the view of health as a right must acknowledge the need for a strong social commitment towards good health.

In a situation where it is utopian to think that the organizations responsible for ensuring human rights will de-neoliberalize themselves, there are two non-exclusive options: on the one hand, actions that influence the systemic repercussions of capitalism, such as the "real utopias" outlined by Erik Olin Wright (2010) or the "seventeen contradictions" by David Harvey (2014); and on the other hand, the realization of the authentic right to health as a first-class human right. The term "social right" deceptively creates a false division between human rights, separating them into "primary" and "secondary" rights. Thus, while "individual rights" such as human dignity and the prohibition of torture are enforced, others considered "social rights," such as access to adequate resources and the elimination of poverty remain only aspirations to be achieved in a more or less distant future.

This separation was cemented in history when on December 16th 1966, the United Nations approved the International Covenant on Economic, Social and Cultural Rights (ICESCR) in which civil and political rights (for example, the prohibition of torture, slavery, the death penalty for minors) were split from economic, social and cultural rights (for example, the right to work, food, housing or health) (Esteban Beltrán 2009). However, poverty (the absence of adequate health, housing, water, food, education or land) should be considered a violation of the human rights as it jeopardizes health and is caused by policies created by specific people and institutions. Let's do the following mental experiment. Imagine that in the case of torture we say: "our objective is to eradicate torture in a decade, and to do

[18]After all if feasibility were a necessary condition for everyone to have any right, "it would be nonsensical to say that everyone has the right to liberty, in view of the difficulty in ensuring the life and liberty of all against transgression."

this we will make a great economic effort and only half of the population will be tortured. (Esteban Beltrán 2009)" If this is unacceptable in the case of torture, why do we accept the same for poverty, the absence of health, work, accessible quality housing, and other social rights that we know directly impact the health of the population?

One fundamental demand of international human rights groups is the necessity of achieving universal access to health as defined, for example, in the third goal "Health and Wellbeing" of the United Nations' *Sustainable Development Goals* (SDGs). However, given that the current widespread approach to realize universal access through basic quality health services, medicine and vaccines, effective and affordable for all (UN SDGs 2016) is based in neoliberal principles,[19] it is no surprise that the immediate antecedent to the SDGs, the Millennium Development Goals approved by 189 countries and set in 2000 for 2015, resulted in an utter failure. And while the causes of this failure, not yet recognized, are multiple and complex, it cannot be ignored that in many countries the objectives were proposed simultaneously with the IMF's Structural Adjustment Programs. Despite the fact that the institutional rhetoric does not admit it and some associates such as Jeffrey Sachs (2005, 2015) continue to receive widespread acclaim, a number of studies clearly show that such policies caused irreparable damages, particularly in African, Southeast Asian, and post-Soviet countries (Stuckler and Basu 2013; Mooney 2012). In sum, serious doubts with respect to the honesty of the proposal are raised when health as a human right is reduced to universal coverage under contemporary neoliberal capitalism, which has not only committed enormous social and health disasters in the "third world" but also through "austerity" policies and other reforms that put at risk the right to health in countries considered models such as Great Britain, Sweden and Spain.

In addition to the gradual destruction of relatively equitable access to quality health services in Western Europe, the current "refugee crisis" reveals the insufficiency and hypocrisy of international and European diplomacy as so many people are abandoned without basic human rights. Deaths at sea, refugee camps, straight deportations, shootings and fences, and the dark relations with Turkey are examples of how many people are treated as "subhuman." In addition, abundant information shows denial of basic human and health rights in immigrant detention centres (the so called CIEs in Spain or "*Centros de Internamiento de Extranjeros*"). As if they were less human than we are. How will human rights, and the right to health, be respected if every day more people lose their citizenship, and those that still formally maintain such rights are being emptied of their republican content?

[19]To expect universal quality coverage in a system that is not public, where insurance companies and the pharmaceutical industry prevail, and where free and quality access is not guaranteed to vulnerable populations as there is no sign of improving their circumstances, seems to be not only practical but also theoretically impossible.

At a moment when the discussion of human rights is in crisis,[20] how do we make possible the right to health?[21] The last decade produced the public health strategy known as "Health in all Policies" whose objective, at least theoretically, is to advance the pursuit of "good health for all" and improve health equity. Through intersectional political, social, economic, scientific and cultural efforts,[22] this strategy attempts to develop systemic responses that establish health as a universal human right which "offers everyone equal opportunities to achieve the highest possible degree of health and wellbeing."[23] Nevertheless, given that people's health fundamentally depends upon material conditions, under rampant and unregulated capitalism fed by accumulation by dispossession, asymmetric power relations and the emergence of large social inequalities, it is utopian to imagine the realization of health as a universal human right. In order to resist the systemic pressures of capitalism, to democratize states and international institutions and to control and regulate the large oligopolies, it is essential to rethink not only the hegemonic discourse of the right to health, but also on the theoretical and practical limitations of implementing the strategy "Health in All Policies".[24]

[20]The failures and contradictions are such that inevitably this discourse sounds empty, instrumental and cynical. Without a doubt there are other examples outside of Europe: the lights and shadows of the peace process in Colombia, the use of drones in imperialist wars, the role of international diplomacy in Libya, Yemen or Syria, the wall and other immigration policies of Trump, the detentions of activists in Russia, Turkey or China, etc.

[21]Sousa Santos points out the weakness of Human Rights as to fully define human dignity, putting forward the substitution of the concept of *utopia* (inspired by Bloch and rethought by Moyn, who considers human rights as the last utopia) for that of *rights*. He also suggests what he terms as "diatopic hermeneutics", one of its consequences being that within the framework of Human Rights there is not consideration for future generations as they still have no duties (climate change and the ecological crisis are paradigmatic examples). See: B. De Sousa Santos. Derechos humanos, democracia y desarrollo. Bogotá: Centro de Estudios de Derecho, Justicia y Sociedad, Dejusticia, 2014. Accessible at: http://www.boaventuradesousasantos.pt/media/Derechos%20Humanos%20Democracia%20y%20Desarrollo.pdf

[22]According to Oxfam's latest report, the inequality of global wealth in 2017 experienced the largest growth in the number of billionaires in history, with more than one new billionaire every 2 days. This huge increase could have ended seven times with extreme poverty worldwide. 82% of all the wealth created last year went to the top 1% wealthiest, and nothing to the lowest 50%. See: Oxfam. Reward Work, Not Wealth: To end the inequality crisis, we must build an economy for ordinary working people, not the rich and powerful. Accessible at: https://policy-practice.oxfam.org.uk/publications/reward-work-not-wealth-to-end-the-inequality-crisis-we-must-build-an-economy-fo-620396

[23]Published by the WHO in December 2015 in a note on health and human rights, 2 months after the launching of the Sustainable Development Goals by the United Nations and inspired also by a 2009 document of the Committee for Economic, Social and Cultural Rights of the United Nations.

[24]To go further, recovering the credibility of the most positive aspects of the Declaration of Human Rights, and reconceptualizing the right to health in positive terms instead of as absence of disease or mere access to medical resources, it is essential that the United Nations, WHO and other international organizations stop applying fundamentally neoliberal ideologies and policies, and that the United States does not boycott minimal agreements as is the case with climate change.

Constitutions, criminal laws, and political practices must change. We must name "social cutbacks" for what they are violations of human rights. We must pressure governments so that proposed cuts do not damage society's most vulnerable populations. It is not possible to be complacent in such a defense; we must be "radical" (approaching the problem from its roots).[25] Wedge a piece of truth in the veil of "post-truth" and pry it open, this should be a fundamental challenge of critical thinking. When we talk about human rights we should at the same time talk about "all rights." To do this, we need systemic political responses. There is no reason why one must choose between eating and being free, between education and life, between health and freedom from torture. It is not only a matter of dignity and justice; the right to health cannot be understood without the other social rights. Only when embodied through political and social struggles and assertion, alongside deep legal and cultural change, can the right to health join the rest as a first level human right.

Acknowledgements We acknowledge the help of Domingo Contreras, Claire Ciraolo and Megan McShea in the translation and edition of some sections of this text. Joan Benach gratefully acknowledges the financial support by ICREA under the ICREA Academia programme.

References

Bourdieu, P. (1977). *Outline of a theory of practice*. Cambridge: Cambridge University Press.

de Rooij, S. R. (2006). *The metabolic consequences of prenatal exposure to the Dutch famine*. Doctoral dissertation. Amsterdam University.

de Sebastián, L. (2009a). *Un planeta de gordos y hambrientos* (p. 186). Barcelona: Ariel.

de Sebastián, L. (2009b). *Un planeta de gordos y hambrientos* (p. 203). Barcelona: Ariel.

de Sousa Santos, B. (2002). Toward a multicultural conception of human rights. In B. Hernández-Truyol (Ed.), *Moral imperialism. A critical anthology* (pp. 39–60). New York: New York University Press.

de Sousa Santos, B. (2014). *Derechos humanos, democracia y desarrollo*. Bogotá: Centro de Estudios de Derecho, Justicia y Sociedad, Dejusticia. Accessed from http://www.boaventuradesousasantos.pt/media/Derechos%20Humanos%20Democracia%20y%20Desarrollo.pdf

De Vogli, R., Kouvonen, A., & Gimeno, D. (2014). The influence of market deregulation on fast food consumption and body mass index: A cross-national time series analysis. *Bulletin of the World Health Organization, 92*, 99–107A.

Domènech, A. (2009, November 29). *Dominación, derecho, propiedad y economía política popular*. Sin Permiso. Accessed from http://www.sinpermiso.info/sites/default/files/textos//dominacion.pdf

Esteban Beltrán. (2009). *Derechos torcidos: Tópicos, medias verdades y mentiras sobre pobreza, política y derechos humanos*. Debate: Barcelona.

Farrés, O. (2016). Ideología y epidemiología. In A. Estany & A. Puyol (Eds.), *Filosofía de la epidemiologia social* (pp. 183–201). Madrid-Méjico: CSIC – Plaza y Valdés.

Federici, S. (2014, May 19). *A woman's body is capitalism's last frontier*. Accessed from http://traficantes.net/noticias-editorial/silvia-federici-de-gira-en-mayo

[25]See: J. C. Monedero. Speech in the 28th regular session of the United Nations Human Rights Council in Geneva about international cooperation in human rights, March 3rd 2015. Accessible at: https://www.youtube.com/watch?v=zYL9aEBATpE; Interview to Esteban Beltrán: "No hay recortes, hay violación de derechos humanos", April 18th 2013. La Opinión A Coruña. Accessible at: http://www.laopinioncoruna.es/contraportada/2013/04/17/esteban-beltran-hay-recortes-hay-violacion-derechos-humanos/712846.html

Garcia, F. (2016, June–August). Dos menos uno, dos. Quién decide el precio de los alimentos. *Boletín Ecos, 35*. Fuhem Ecosocial. Accessed from https://www.fuhem.es/media/cdv/file/biblioteca/Boletin_ECOS/35/quien-decide-precio-alimentos_F_GARCIA.pdf
Harvey, D. (2014). *Seventeen contradictions and the end of capitalism*. New York: Oxford University Press.
Herrero, Y. (2017, July 12). Speech at the "Jornadas de Debate Feminista 2017". Montevideo, Uruguay. Accessed from https://www.youtube.com/watch?v=iHjxuJ8ks_Q
Joseph, J. (2015). *The trouble with twin studies: A reassessment of twin research in the social and behavioral sciences*. New York: Routledge.
Krieger, N. (2001). Glossary of social epidemiology. *Journal of Epidemiology and Community Health, 55*, 693–700.
Krieger, N. (2013). History, biology, and health inequities: Emergent embodied phenotypes and the illustrative case of the breast cancer estrogen receptor. *American Journal of Public Health, 103* (1), 22–27.
Krieger, N., & Smith, G. D. (2016). The tale wagged by the DAG: Broadening the scope of causal inference and explanation for epidemiology. *International Journal of Epidemiology, 45*, 1787–1808.
Lewontin, R. C. (1991). *Biology as ideology*. New York: Harper Collins.
Lewontin, R., & Levins, R. (2007). *Biology under the influence. Dialectical essays on ecology, agriculture, and health*. New York: Monthly Review Press.
Lewontin, R. C., Rose, S., & Kamin, L. J. (1984). *Not in our genes. Biology, ideology and human nature*. New York: Penguin/Pantheon Books.
Mann, J. M., Gostin, L., Gruskin, S., Brennan, T., Lazaarini, Z., & Fineberg, H. V. (1994). Health and human rights. *Health and Human Rights, 1*, 6–23.
Marx, K., & Engels, F. (1978). The German ideology. In R. C. Tucker (Ed.), *From the Marx-Engels reader* (2nd ed.). New York: W.W. Norton.
Monedero, J. C. (2015, March 3). Speech in the 28th regular session of the United Nations Human Rights Council in Geneva about international cooperation in human rights. Accessed from https://www.youtube.com/watch?v=zYL9aEBATpE
Mooney, G. (2012). Neoliberalism is bad for our health. *International Journal of Health Services, 42*, 383–401.
Moss, M. (2003). *Salt sugar fat: NY Times Reporter Michael Moss on how the food giants hooked America on junk food*. A Goodman. Interview to M Moss. Democracy Now.
Nestle, M. (2002). *Food politics* (p. 27). Berkeley: University California press.
Riechmann, J. (2003). *Cuidar la T(t)ierra. Políticas agrarias y alimentarias sostenibles para entrar en el siglo XXI*. Barcelona, Icaria.
Sachs, J. (2005). *The end of poverty: How we can make it happen in our lifetime*. London: Penguin.
Sachs, J. (2015). *The age of sustainable development*. New York: Columbia University Press.
Schuldt, J. (2004, July 28). Somos pobres porque somos ricos. Rebelión. Accessed from http://www.rebelion.org/noticias/2004/7/2579.pdf
Sen, A. (2008). Why and how is health a human right? *The Lancet, 372*, 2010.
Stuckler, D., & Basu, S. (2013). *The body economic: Why austerity kills*. New York: Basic Books (AZ).
UN SDGs. (2016). *Goal 3. Good health and well-being. Ensure healthy lives and promote well-being for all at all ages*. http://www.un.org/sustainabledevelopment/health/
Vallverdú, J. (2016). Causalidad y epidemiología. In A. Estany & A. Puyol (Eds.), *Filosofía de la epidemiologia social* (pp. 99–118). Madrid-Méjico: CSIC – Plaza y Valdés.
Varoufakis, Y. (2017). *Talking to my daughter about the economy*. London: Bodley Head.
Vasak, K. (1982). *The international dimensions of human rights* (Vol. 1). Westport, CT: Greenwood Press.
Wallerstein, I. (1983). *Historical capitalism*. London: Verso Editions.
WHO. (2016, June). *Obesity and overweight*. Fact sheet num 311. Accessed from http://www.who.int/mediacentre/factsheets/fs311/en/
Wright, E. O. (2010). *Envisioning real utopias*. London: Verso.

Ethics and Public Health Policies

Ildefonso Hernández-Aguado, Elisa Chilet-Rosell, and María Pastor-Valero

Abstract Ideological differences between the individual and population focus spans all public health actions from research to policy decisions. Growing interest in applying ethics to public health has strengthened the theoretical basis for public health to address the fundamental causes of disease. It bestows on governments a key role in improving population health, and underlines the participation and responsibility of all levels of society. The debate on whether public health research and practice should engage in political action touches on several issues. We focus here on two main functions of public health: governance and advocacy. The role of the government in promoting health and reducing inequalities is challenged in at least three areas: the capacity of public administrations; the engagement of public–private interactions; and, the use of law to improve population health. An adverse global political and ideological environment causes the loss of governments' capacity to protect health, promotes risky interactions with transnational corporations to promote health and hampers the use of law as a public health tool. In this context, public health advocacy emerges as a key action to advance population health, although there are several visions on which type of advocacy better integrates public heath principles and values. The contributions of the humanities could benefit public health research and practice and help public health to face its present and future challenges.

Keywords Health policy · Governance · Advocacy · Paternalism

I. Hernández-Aguado (✉) · E. Chilet-Rosell · M. Pastor-Valero
Department of Public Health, History of Science and Gynaecology, School of Medicine, University Miguel Hernández de Elche, San Juan de Alicante, Spain

Ciber de Epidemiología y Salud Pública (CIBERESP), Madrid, Spain
e-mail: ihernandez@umh.es; echilet@umh.es; mpastor@umh.es

J. Vallverdú et al. (eds.), *Philosophical and Methodological Debates in Public Health*, https://doi.org/10.1007/978-3-030-28626-2_13

Introduction

Public health prospers when society considers it as a public good. When the community lacks a sense of the common good, public health action diminishes and is unnoticed under the dominance of the prevailing biomedical and individual approach to health issues. Public health has been defined as 'the art and science of preventing disease, prolonging life and promoting health through the organized efforts of society' (Acheson 1988). However, these organized efforts of society become diverse realities according to political, ideological social and scientific factors. It is well established that there is an ideological tension between individual and population focus that spans all public health actions from research to policy decisions.

The response to health problems is primarily grounded in the biomedical model that limits the social response to those actions that are provided through healthcare services. The biomedical approach attributes health problems to individual circumstances and behaviours and consequently the solutions tend to be of an individual nature. Prevention usually becomes limited to individual centred activities, such as medical counselling to modify health related behaviours, or screening for early diagnosis. This narrow view of public health fails to identify the causes of causes of health and disease and to act accordingly.

Originally, public health was conceived to distinguish actions that governments and societies—as opposed to private individuals—should take to preserve and protect the people's health. Founded in social justice, there is a strand of public health that is committed to the reduction of social inequalities in health and that is set in collective action 'to make the world a better place, free of misery, inequity, and preventable suffering, a world in which we all can live, love, work, play, ail, and die with our dignity intact and our humanity cherished' (Krieger and Birn 1998).

In addition to the ideological tension in public health, recent decades have witnessed a remarkable accumulation of knowledge recognizing the social nature of health problems together with theoretical developments and calls for actions. From the Declaration of Alma-Ata (WHO 1978) and the Ottawa Charter for Health Promotion (WHO 1986) to the Report of the Commission on Social Determinants of the World Health Organization (Commission on Social Determinants of Health 2008), a broad view of public health founded on social justice and respect for human rights has developed. Thus, equity in health is an expression of social justice and recognizes that the enjoyment of the highest attainable standard of health is one of the fundamental rights of every human being without distinction of race, religion, political belief, gender, economic or social condition. Jonathan Mann conceived human rights and ethics as central to public health action and stated that if its mission is to assure the social conditions for the population's health, then public health must address the fundamental determinants of wellbeing (Mann 1997). The report of the Commission on Social Determinants of Health proposes public policies at all government levels to address the radical causes of ill health.

Current ethical frameworks are usually based on the original notion of public health. The Principles of the Ethical Practice of Public Health of the American Public Health Association are based on values and beliefs that share the above perspective and state that humans have a right to the resources necessary for health and that humans are inherently social and interdependent (Thomas et al. 2002). The Principles affirm that public health should address mainly the fundamental causes of disease and requirements for health, in order to prevent adverse health outcomes and that public health should advocate and work for the empowerment of disenfranchised community members, in order to ensure that the basic resources and conditions necessary for health are accessible to all. Likewise, the Ethical Framework for Policy of the Nuffield Council of Bioethics proposes the use of regulation to reduce the causes of ill health by ensuring environmental conditions that sustain good health and by reducing health inequalities (Nuffield Council of Bioethics 2007).

There has been a growing interest in the application of ethics to public health as illustrated by the publication of several books describing numerous practical cases spanning the globe (UK Health Forum 2018; Barrett et al. 2016; Fundació Víctor Grífols i Lucas 2013). Also, the use of law to improve population health is now a target of renewed interest from the academic field and health institutions. The publication of the book 'Advancing the right to health: the vital role of law', a result of the collaboration between the WHO, the Development Law Organization, the O'Neill Institute for National and Global Health Law and the Sydney Law School is good news and offers a wide range of strategies by which law can favour population health (WHO et al. 2017; Gostin et al. 2017). This progress is strengthening the theoretical basis for the implementation of a vision of public health that addresses the fundamental causes of disease, that bestows on governments a key role in improving population health and underlines the participation and responsibility of all levels of society. However, all the advances described have flourished at a time when, in the words of Krieger, the very notion of public health as social good is challenged by profit-driven agendas and when the role of the state in fostering human welfare is constantly questioned (Krieger and Birn 1998). The goal remains: a just and sustainable world in which we and every being on this planet may truly thrive (Krieger 2015). However, present public health faces an even worse political and social environment with problems such as income and wealth inequalities or climate change becoming more pronounced. The same idea of public health as a public good and not as a commodity is under threat. We believed that good science and good evidence would influence policies favouring the public health vision but this idea was more wishful thinking based on an unsatisfactory understanding of the political and policy processes. Public health researchers are now coming closer to political science and are realizing that a change of approach is needed if we are to have more political incidence. There are innovative proposals to shape public health policy, such as the development of narratives linked to evidence based ideas (Smith 2013).

Whether public health research and practice should engage in political action or not, is subject to a debate that covers a wide range of topics. In order to consider the diverse issues and the role of ethics in these debates on health policy, we will focus on two of the four key areas described by the World Federation of Public Health

Associations in the Global Charter for the Public's Health: governance and advocacy (Lomazzi 2016).

Governance

The role of the government in promoting health and reducing inequalities is challenged by different ideological positions, and the resulting tension can be envisaged in at least three areas: the capacity of public administrations; the engagement in public private interactions; and, the use of law to improve population health.

Regarding administration capacity, a public health policy in accordance with the aforementioned ethical frameworks requires leadership and good governance. Cuellar, in his analysis of regulatory capture and public health, shows how the positive features of an agency such as the prestigious US Centre for Disease Control and Prevention (CDC) has engendered a reputational capital that places it in an influential position which is shaping public health policy (Cuellar 2014). The ability, credibility and leadership of the public administration and its agencies should, therefore, be recognized. Using several examples, he points out that the CDC's low number of conflicts of interest and its scientific capacity confer the agency with the public legitimacy to influence policies and regulations and to give credit to the government's political innovations regarding health. The CDC and other influential health agencies are essential for national and global health governance and they have the advantage of having a reservoir of legitimacy reflecting the broad association of modern health policy with scientific expertise in the public sphere (Cuellar 2014).

Unfortunately, there has been a series of events that are threatening the independence of health agencies. The considerable credibility of the CDC could begin to teeter as a result of the conflicts of interest described in the British Medical Journal (Lenzer 2015) or because of its refusal to carry out research on gun violence in order not to lose funds (Rubin 2016). The fissures appearing in public confidence in the agency could deepen due to its director's recent resignation because her financial conflicts of interest prevented her from doing her job properly (Sun 2018). The health related European agencies are not in a better position. The European Food Safety Authority and European Medicines Agency are among the most often mentioned European agencies with regard to conflicts of interest and regulatory capture (Robinson et al. 2013; Cohen 2012; Diamandouros 2011; European Parliament 2012). The WHO has also been suspected of involvement in some conflicting situations (Cohen and Carter 2010). Although all these institutions are performing appropriately in many instances, the erosion of public confidence in them could have marked and long lasting negative effects in governance, as it takes time to recover credibility.

The National Organization of Transplants in Spain (ONT) illustrates how public organizations can assume their responsibilities without succumbing to undue influences of private interests. The Deutsche Knochenmarkspenderdatei (DKMS) company tried to introduce profit-driven activities in the search for donors; the German company moved a team to Spain to carry out a campaign to find donors which clashed with the principles of the Spanish organization, established in the National Health System. This company was targeting the ONT, but the media was also a target of influence: they received invitations to defend this search for donors. Some media reproduced the communications of the German company whose strategy was to turn patients in the transplant list against the ONT and the Ministry. The company used the services of a law firm to litigate against the Ministry and to influence the media. Finally, the ONT got DKMS not to act in Spain, arguing that the DKMV's commercial offer conflicted the established order to get an organ transplantation based on the needs of patients and put the equity of the Spanish National Health System at risk. Many positive features of the organization (ONT) such as free provision, equity and success (world leader in organ donation and transplantation) ensures governance and reduces its vulnerability to external influences.

Institutional quality enables stronger leadership from the public administration and respect from the public. But the high stakes involved in public health decisions attract the participation of multiple political actors. In this policy space some transnational corporations are becoming crucial actors that explain some governance failures and regulatory capture (Mindell et al. 2012). Most of the questioning on the independency of agencies is due to interferences by private companies related to health or disease. The potential dissipation of the capital of trust of health agencies is due in part to the political environment that undermines the capacity of the states and of public administrations. The need for additional funding in view of the budget constraints of public origin has conducted public health agencies to arrangements based in private financing that threats their independence.

Due to similar reasons, the ability of public organizations to execute their responsibilities in the search for public interest and maintaining independence from narrowly motivated private interests is becoming a difficult enterprise. Current solutions include public private arrangements and some governments have introduced partnerships with corporations as a key element of health strategies (Ginn 2011). Paradoxically, debates are not centred on the causes of state shrinking that lead to budgetary restrictions and jeopardize the capacity of governments to comply with their responsibilities, but rather whether interactions with private companies in policy making are appropriate.

There is a growing interest in using public–private partnerships to address health-related issues. In fact, some researchers and interested parties argue that engagement in partnerships is an ineluctable path towards improvements in population health. This interest began in the last century when neoliberal ideologies influenced public policy and influential international organisations acknowledged and championed a greater role for the private sector. The nineties see clear development of PPPs in the United Nations, including the WHO, of which the causes and landmarks have been well described by (Richter 2004). The range of public private interaction in purpose, design and composition is broad. There are arrangements to provide drugs for neglected diseases or vaccines in low income countries but also agreements with transnational corporations of alcohol producers to promote health. The halo of theoretical success—still not appropriately evaluated—of some arrangements used in global health together with a favourable intellectual environment could have supported the emergence of public–private partnerships in health promotion (Hernández-Aguado and Zaragoza 2016).

The question of governments engaging with corporations in order to promote health is a central issue in public health today and should be the object of careful consideration. The debate is focused on whether these interactions actually promote health and what conditions governments should impose before engaging with industries.

The arguments of the supporters of public–private partnerships for health can be summarized in the following points (Hernández-Aguado and Zaragoza 2016): (1) the immense threats to health cannot be tackled by governments alone; (2) public–private partnerships enrich the capacity, quality and scope of public health services; (3) industries can benefit from public health service expertise; (4) partnerships help to put health in all policies and improve self-regulation; and (5) public private interactions reduce unhealthful products and improve the quality of products. Authors critical of this strategy suggest that: (1) alliances between public health and the private sector of which the products or services are unhealthful have inherent conflicts of interest that cannot be reconciled; (2) collaborations in health promotion confers legitimacy and credibility on industries that produce disease related products and can damage the credibility of public health institutions; (3) public private partnerships capture institutions (UN agencies, governments, etc.), regulatory bodies and science; (4) the lack of evidence on effectivity calls for the application of precautionary principle; and (5) objectives of collaborations contradict public health priorities. Some authors recommend conditions for engaging in public private partnerships that are related to governance, type of industry and role of the partner in policy (Hernández-Aguado and Zaragoza 2016). Galea and McKee (2014) propose five tests that public policy makers may wish to apply before engaging in such a public–private partnership. The first one refers to a critical issue: are the core products and services provided by the corporation health enhancing or health damaging? Indeed, most partnerships in health promotion involve corporations that have been considered vectors of disease (Gilmore et al. 2011).

We refer here to collaborations in health promotion where it is difficult to find any example of effectivity. But, there are also concerns about public private interactions for the donation of drugs and medical devices which may not be appropriate to the needs of countries, and may distort national priorities in their healthcare services (van Dijk et al. 2011). Behind the support of public–private partnerships are the same ideas that encourage the reduction of the size of the state or that welcomes the spread of philanthropy in global health. We believe that a critical analysis of the appropriateness of public private engagements in health is required, it is also essential to examine the 'philanthrocapitalism' phenomenon and its effect on global health governance (Clark and McGoey 2016). The study of social determinants of health must not obviate the role of transnational corporations and their related organizations as structural causes of health and disease. However, we cannot forget that the ethos and nature of these trends are closely related to previous forms of charity and to the expectations of the market as provider of the common good (McGoey 2012). The debate is indeed not only on the role of governments in promoting health and reducing inequalities but also on the role of the state in modern societies. Regarding public private collaborations, public health practitioners and researchers now have numerous practice based cases available to guide them (UK Health Forum 2018).

Most of the cases that analyse public private interactions conclude that unless governments apply regulation to control unhealthy commodities, it will not be possible to reduce the diseases related to these products. The use of law to protect health reaches far beyond the regulation of harmful foods and drinks (and of course tobacco) as is well illustrated by the comprehensive work recently published by the WHO (WHO et al. 2017). The health, social and economic benefits brought by regulation inside the health field (food and water security, tobacco, alcohol, etc.) and outside (banning of leaded petrol, chemical security, etc.) are enormous, and perhaps inadequately communicated to the public. In spite of this wide beneficial quality of these laws, any attempt to use regulation to protect health is strongly resisted by private interests.

Opposition by industries to health protection resembles a huge remora that has pestered public health for more than a century. Michaels has described in his book 'Doubt is their product' how industry has manufactured and magnified scientific uncertainty, and influenced policy decisions to the advantage of polluters and the manufacturers of dangerous products (Michaels 2008). Wiist has suggested that food companies may be replicating several of the tactics used by tobacco companies to promote a positive image and influence research as part of an effort to reduce the prospects of public regulation and taxation (Wiist 2011). In the examples of what Wiist calls 'the corporate playbook', he describes the following public relations tactics: emphasizing consumers' personal responsibility, moderation, free choice, and pleasure; using 'government' versus 'personal freedom and civil liberties' and 'Get government off our backs' arguments; and vilifying critics, health advocates and public health scientists as 'health police' or 'fascists' and accusing them of seeking to impose a 'nanny state'.

Paternalism and public health intervention is a classic and permanent debate. Principles of public health ethics address the issue when acknowledging that public health programmes and policies should incorporate a variety of approaches that anticipate and respect diverse values, beliefs, and cultures, and also achieve community health in a way that respects the rights of individuals in the community (Thomas et al. 2002). Interventions should identify the burdens or harms that might exist and seek to minimise them or search for alternatives. The so-called 'intervention ladder' proposed by the Nuffield Council of Bioethics showing the incremental degree of interference in the personal sphere could be suitable to reduce risks (Nuffield Council of Bioethics 2007). All public health actions could cause burdens such as: risks to privacy and confidentiality, risks to liberty and self-determination, and risks to justice. The likelihood of each risk depends on the particular type of public health strategy or programme (Kass 2001). Consequently, public health policies must examine comprehensively the potential harm; however, most debates are not related to the degree of paternalism inherent to the specific public health measure but to the role played by the interested actors and their power.

Health protection implemented by the state is far wider than usually perceived. In our daily life we come across public health interventions that involve some degree of that paternalism which produces high collective benefits and low individual burdens. We refer to regulations that assure safety of drugs, high standards of healthcare professionals, safe food and water, air quality, absence of toxicity in work and home environments, etc. In the words of Gostin and Gostin (2009): 'Public health paternalism that markedly improves health and well-being within the population offers a 'broader freedom'. This term is used advisedly to mean that when people have better opportunities for health and longevity, and live in more vibrant, productive communities, they have enhanced prospects for life and a wider range of choices for now and into the future'. Most public health interventions are not called into question for their risk to freedom, either to some part or to the whole population. Some groups that argue against taxation of unhealthy commodities or against limits to the use of individual vehicles in some city areas are also opposed to the right of all people to a dignified death which respects their personal and cultural values. Some that reject regulations that would even be accepted applying the 'harm principle' of John S. Mill, are willing to accept limits to sexual freedom.

We do not believe that the question is paternalism against freedom at the individual level, despite its relevance in ethical discussions. The relevant framework is the community, understood as an interdependent social body where the limits of law are basic liberties, human rights and effective social justice. If we accept this framework, then the main concerns of public health will not be the level of taxes on some products—that are ineluctable—but rather the guilty inaction of governments in failing to protect their populations. When human rights are infringed in wealthy societies, debates on paternalism should focus on whether to limit the freedom of those in power and to force them to satisfy the essential needs of their populations. If the goal of public health were power redistribution and respect for the rights of people in order to improve population health, then there would be less need for regulation of the majority of people and more laws centred on those with power, both at public and private level. Under these conditions, the risks to liberty would be restricted to those with an excess of power and consequently there would be less concern for loss of freedom (Hernandez Aguado 2017).

In summary, and regarding to governance issues, the capacity of public administrations, the engagement in public private interactions, and the use of law to improve population health are all part of the challenges to public health. The wide range of undue influences that are exerted on policy-makers and executive officials are converting the public health policy making environment into a complex field (Hernández-Aguado and Chilet-Rosell 2018). Gollust et al. (2008) also described how political issues also incorporate ethical challenges to public health practitioners particularly in four main areas: public health agenda-setting, political pressures, political conflicts with best practices, and the scope of public health practice. Consequently, and in order to address these trials, well equipped decision-makers and practitioners with a sound training in ethics and good governance are needed.

Advocacy

Ethics frameworks include the need for community participation in the formulation and implementation of public health policies. The frameworks put forward the principle that each person in a community be given the opportunity to contribute to public discourse and insist that it is necessary to make procedural justice arrangements (such as democratic decision-making procedures) to obtain the assent of communities in the development of public health interventions. Present public policy making is characterised by democratic imbalance in which some policy actors exhibit greater capacity of participation and influence on decisions taken. This state of affairs is not new and explains why advocacy has proven an essential recourse to promote healthy public policies. The history of public health would have been different if some researchers and practitioners had not developed sensible health advocacy strategies. The experience with tobacco and other causes of disease show the inestimable role of public health advocacy in securing public and political support for health protection legislation and policies. Australian tobacco control owes its success to health advocates. It is an example of good practice which is widely acknowledged and enshrined in WHO policy documents (Chapman and Wakefield 2001). This type of health advocacy that aims to protect population health and mainly the vulnerable was defined by Carlisle as 'representational advocacy' (Carlisle 2000). Thus far, 'representational advocacy' has prevailed in public health and, to a certain extent, could be considered as a form of lobby by which practitioners and researchers seek to influence decision-makers and the public for the advance of public health policies. Smith and Stewart argue that this approach to advocacy can be criticised as technocratic and elitist, although they acknowledge its value to put under-represented public health concerns higher up the policy agenda (Smith and Stewart 2017). They propose 'facilitational advocacy'—coined by Carlisle for the type of advocacy that empowers the disadvantaged—as an alternative. According to Smith and Stewart, 'facilitational' advocacy' is a more democratic approach that listens and works with communities and its members whose voices are under-represented in research and policy debates (Smith and Stewart 2017).

There is no doubt that a health advocacy that contributes to a more intense and democratic participation of the public in shaping research and policy fits appropriately with public health ethical values. Nevertheless, there are solid reasons to support both types of advocacy described. The space of policy decision has become complex. There is a clear expansion on the horizontal axis with an increasing number of involved actors such as transnational corporations, non-governmental organizations (NGOs), the so-called global civil society, etc. On the other hand, nation states are progressively losing power on the vertical axis both upwards to supranational bodies and institutions and downwards to regional and local authorities. Any comprehensive analysis of the social determinants of health should pay attention to all decision levels. Nutrition related problems, for example, are the result of policies implemented at global, national, regional and local levels with complex interactions among a myriad of actors and factors. In the European Union, important decisions on nutrition are taken at the supranational level in an intricate process that involves different institutions with varied roles and power: mainly the European Commission, the European Parliament, and the Council of the EU. The ability of NGOs committed to public health values and goals to influence European policies is meagre compared to private interests. One relevant decision was related to new European legislation on food labelling—determining what nutritional information should be displayed on the packaging of items such as snacks, soft drinks and ready-meals. Public health advocates and other consumer organizations favoured the traffic-light label system which allows a much easier understanding for the public. According to the Corporate European Observatory (a watchdog organization), the Confederation of the Food and Drink Industries of the EU (CIAA) spent 1 billion euros opposing proposals for front-of-pack 'traffic light' labels in favour of a system based on guideline daily amounts (Corporate Europe Observatory 2010). There was a scientific discussion on the capacity of industry and NGO lobbies at the European Parliament (Kurzer and Cooper 2013; Hoff et al. 2016), but the imbalance of power and activities implemented at all EU power levels was obvious and the final results were favourable to the industry. This case also illustrates the contradictions of the food and drink industry when it puts forward the argument of paternalism against regulation as a threat to freedom of choice, while at the same time obscuring the information to consumers by countering traffic-light labels, front of packaging labels and large printing. When uneven capacity of democratic participation is so huge, only a strong 'representational advocacy' capacity could strive for a more balanced picture. Decisions on food policies at multilateral institutions such as the WHO and the World Trade Organization show a similar complexity to European processes and also share the need to strengthen the public health voice. 'Representational advocacy' has an essential role when the distance between the policy making process and the public health community is greater, but even on local and regional environments there are examples of success.

While advocacy is increasingly viewed as a nuclear part of public health functions, there is still a substantial debate on its precise role. Smith and Stewart portray the variety of perspectives of past and current debates on advocacy in the public health field which is mainly focused on what advocacy is and who ought to undertake it. They reviewed the scientific literature on advocacy and undertook qualitative research involving 147 professionals from diverse public health settings and concluded that 'divergent ways of thinking about advocacy speak to a more fundamental challenge regarding the role of the public in research, policy and practice and the activities that connect these various strands of public health research' (Smith and Stewart 2017).

We believe that the role of the public is becoming a central issue in modern public health. We need intensive research on the most effective procedures to involve the population in the shaping and implementation of public health interventions. This should be parallel to improvements and expansion in 'representational advocacy' where it is necessary and effective. On the other hand, we feel that discussing the individual roles of academics and practitioners in advocacy is a mistake. Advocacy, as any other public health activity, should be a shared initiative based on building alliances. For the authors, the question is not so much whether a researcher or practitioner should take part in advocacy, but rather whether he or she should be involved with others to defend population health on the grounds of agreed values as described in the ethical frameworks mentioned above.

Concluding Remarks

The approaches to population health are traditionally grounded in the advances of biomedical sciences. Influences of biomedicine have resulted in undeniable benefits to public health and part of its public legitimacy and credibility is based on the scientific contributions from medicine. However, the biomedical perspective promotes health care based individual responses to population health issues. There is enough science-based knowledge to support population and community approaches through the implementation of public policies, both inside and outside the health field, addressing the radical causes of disease. From an ethical perspective, any public health interventions should consider the requirement to ensure environmental conditions that sustain good health and reduce health inequalities. Conceiving public health as a collective action with wide societal values such as human rights and social justice can introduce difficulties when it comes to setting targets or achieving goals. The global political and ideological environment erodes the idea of shared values and the consideration of public health action as public good. In this context, public health is drawn into debates on what should the role of governments be in protecting health and on the risks that population oriented measures might bear to individuals. The debates on paternalism that some public health interventions engender or discussions on the role of private actors in policymaking could be obscuring more essential debates. Public health researchers and practitioners should begin to consider the primary causes that drive neglect of sound public health policies and reflect on whether contributions from the field of humanities—such as philosophy and politics—could benefit public health training and research by increasing the science base and building arguments to defend population heath.

References

Acheson, D. (1988). *Public health in England. The report of the committee of inquiry into the future development of the public health function*. London: H.M. Stationery Office.

Barrett, D. H., Ortmann, L. W., Dawson, A., Saenz, C., Reis, A., & Bolan, G. (Eds). (2016). *Public health ethics: Cases spanning the globe*. Switzerland: Springer. Accessed February 8, 2008, from http://www.springer.com/us/book/9783319238463

Carlisle, S. (2000). Health promotion, advocacy and health inequalities: A conceptual framework. *Health Promotion International, 15*, 369e376. https://doi.org/10.1093/heapro/15.4.369.

Chapman, S., & Wakefield, M. (2001). Tobacco control advocacy in Australia: Reflections on 30 years of progress. *Health Education & Behavior, 28*(3), 274–289. https://doi.org/10.1177/109019810102800303.
Clark, J., & McGoey, L. (2016). The black box warning on philanthrocapitalism. *The Lancet, 388*(10059), 2457–2459. https://doi.org/10.1016/S0140-6736(16)32184-5.
Cohen, D. (2012). European Medicines Agency is to tighten up on advisers' conflicts of interest. *British Medical Journal, 344*, e2388. https://doi.org/10.1136/bmj.e2388.
Cohen, D., & Carter, P. (2010). Conflicts of interest. WHO and the pandemic flu "conspiracies". *British Medical Journal, 340*, c2912. https://doi.org/10.1136/bmj.c2912.
Commission on Social Determinants of Health. (2008). *Closing the gap in a generation: Health equity through action on the social determinants of health: Commission on Social Determinants of Health final report*. Ginebra: World Health Organization. Accessed February 8, 2018, from http://www.who.int/social_determinants/thecommission/finalreport/en/
Corporate Europe Observatory. (2010). *A red light for consumer information*. Accessed February 8, 2018, from http://corporateeurope.org/sites/default/files/sites/default/files/files/article/ceo-food-labelling.pdf
Cuellar, M. F. (2014). Coalitions, autonomy & regulatory bargains in public health law. In D. Carpenter & D. A. Moss (Eds.), *Preventing regulatory capture special interest influence and how to limit it. The Tobin Project* (pp. 326–362). New York: Cambridge University Press.
Diamandouros, P. N. (2011). *Draft recommendations of the European Ombudsman in his inquiry into complaint 775/2010/ANA against the European Food Safety Authority*. Accessed February 8, 2018, from http://www.ombudsman.europa.eu/en/cases/recommendation.faces/en/11089/html.bookmark
European Parliament. (2012). *Decision of 10 May 2012 on discharge in respect of the implementation of the budget of the European Food Safety Authority for the financial year 2010*. (C7-028672011/2226(DEC). Accessed February 8, 2018, from http://www.europarl.europa.eu/sides/getDoc.do?type=REPORT&reference=A7-2012-0299&language=EN
Fundació Víctor Grífols i Lucas. (2013). *Casos prácticos de ética y salud pública N° 29*. Barcelona: Fundación Grífols.
Galea, G., & McKee, M. (2014). Public–private partnerships with large corporations: Setting the ground rules for better health. *Health Policy, 115*, 138–140. https://doi.org/10.1016/j.healthpol.2014.02.003.
Gilmore, A. B., Savell, E., & Collin, J. (2011). Public health, corporations and the new responsibility deal: Promoting partnerships with vectors of disease? *Journal of Public Health (Oxford, England), 33*, 2–4. https://doi.org/10.1093/pubmed/fdr008.
Ginn, S. (2011). "Responsibility deal" is unlikely to produce health benefits, says report. *British Medical Journal, 343*, d5791. https://doi.org/10.1136/bmj.d5791.
Gollust, S. E., Baum, N. M., & Jacobson, P. D. (2008). Politics and public health ethics in practice: Right and left meet right and wrong. *Public Health Management Practice, 14*(4), 340–347. https://doi.org/10.1097/01.PHH.0000324561.42039.47.
Gostin, L. O., & Gostin, K. G. (2009). A broader liberty: J.S. Mill, paternalism and the public's health. *Public Health, 123*(3), 214–221. https://doi.org/10.1016/j.puhe.2008.12.024.
Gostin, L. O., Magnusson, R. S., Krech, R., et al. (2017). Advancing the right to health-the vital role of law. *American Journal of Public Health, 107*(11), 1755–1756. https://doi.org/10.2105/AJPH.2017.304077.
Hernández Aguado, I. (2017). La libre opción individual y la salud. Paternalismo directivo o irresponsabilidad pública ante el sufrimiento y la desigualdad. *Folia Humanística, 5*, 1–21.
Hernández-Aguado, I., & Chilet-Rosell, E. (2018). Pathways of undue influence in health policy-making: A main actor's perspective. *Journal of Epidemiology and Community Health, 72*(2), 154–159. https://doi.org/10.1136/jech-2017-209677.
Hernández-Aguado, I., & Zaragoza, G. A. (2016). Support of public-private partnerships in health promotion and conflicts of interest. *British Medical Journal Open, 6*(4), e009342. https://doi.org/10.1136/bmjopen-2015-009342.

Hoff, A., Lelieveldt, H., & van der Does, R. (2016). A biased rapporteur or politics as usual? Reassessing the balance of interests in the EU food information labelling case. *Journal of European Public Policy, 23*(2), 296–313. https://doi.org/10.1080/13501763.2015.1051095.

Kass, N. E. (2001). An ethics framework for public health. *American Journal of Public Health, 91*(11), 1776–1782. https://doi.org/10.2105/AJPH.91.11.1776.

Krieger, N. (2015). Public health, embodied history, and social justice: Looking forward. *International Journal of Health Services, 45*(4), 587–600. https://doi.org/10.1177/0020731415595549.

Krieger, N., & Birn, A. E. (1998). A vision of social justice as the foundation of public health: Commemorating 150 years of the spirit of 1848. *American Journal of Public Health, 88*(11), 1603–1606. https://doi.org/10.2105/AJPH.88.11.1603.

Kurzer, P., & Cooper, A. (2013). Biased or not? Organized interests and the case of EU food information labelling. *Journal of European Public Policy, 20*(5), 722–740. https://doi.org/10.1080/13501763.2012.751703.

Lenzer, J. (2015). Centers for disease control and prevention: Protecting the private good? *British Medical Journal, 350*, h2362. https://doi.org/10.1136/bmj.h2362.

Lomazzi, M. A. (2016). Global Charter for the Public's Health – The public health system: Role, functions, competencies and education requirements. *European Journal of Public Health, 26*(2), 210–212. https://doi.org/10.1093/eurpub/ckw011.

Mann, J. M. (1997). Health and human rights: If not now, when? *Health and Human Rights, 2*(3), 113–120. https://doi.org/10.2105/AJPH.96.11.1940.

McGoey, L. (2012). Philanthrocapitalism and its critic. *Poetics, 40*, 185–199. https://doi.org/10.1016/j.poetic.2012.02.006.

Michaels, D. (2008). *Doubt is their product: How industry's assault on science threatens your health.* Oxford: Oxford University Press.

Mindell, J. S., Reynolds, L., Cohen, D. L., & McKee, M. (2012). All in this together: The corporate capture of public health. *British Medical Journal, 345*, e8082. https://doi.org/10.1136/bmj.e8082.

Nuffield Council on Bioethics. (2007). *Public health: Ethical issues*. London: Nuffield Council on Bioethics. Accessed February 8, 2018, from http://nuffieldbioethics.org/wp-content/uploads/2014/07/Public-health-ethical-issues.pdf

Richter, J. (2004). Public–private partnerships for health: A trend with no alternatives? *Development, 47*, 43–48. https://doi.org/10.1057/palgrave.development.1100043.

Robinson, C., Holland, N., Leloup, D., & Muilerman, H. (2013). Conflicts of interest at the European Food Safety Authority erode public confidence. *Journal of Epidemiology and Community Health, 67*(9), 717–720. https://doi.org/10.1136/jech-2012-202185.

Rubin, R. (2016). Tale of 2 agencies: CDC avoids gun violence research but NIH funds it. *Journal of the American Medical Association, 315*(16), 1689–1691. https://doi.org/10.1001/jama.2016.1707.

Smith, K. (2013). *Beyond evidence-based policy in public health. The interplay of ideas*. New York: Palgrave Macmillan.

Smith, K. E., & Stewart, E. A. (2017). Academic advocacy in public health: Disciplinary 'duty' or political 'propaganda'? *Social Science & Medicine, 189*, 35e43.

Sun, L. H. (2018). CDC director resigns because of conflicts over financial interests. *The Washington Post*. Accessed February 8, 2018, from https://www.washingtonpost.com/news/to-your-health/wp/2018/01/31/cdc-director-resigns-because-of-conflicts-over-financial-interests/?utm_term=.66012d292d1e

Thomas, J. C., Sage, M., Dillenberg, J., & Guillory, V. J. (2002). A code of ethics for public health. *American Journal of Public Health, 92*(7), 1057–1059.

UK Health Forum. (2018). *Public health and the food and drinks industry: The governance and ethics of interaction. Lessons from research, policy and practice*. London: UK Health Forum. Accessed February 8, 2018, from http://www.ukhealthforum.org.uk/prevention/pie/?entryid43=58305

van Dijk, D. P., Dinant, G., & Jacobs, J. A. (2011). Inappropriate drug donations: What has happened since the 1999 WHO guidelines? *Education for Health (Abingdon), 24*(2), 462.

WHO. (1986). *The Ottawa Charter for Health Promotion. First International Conference on Health Promotion, Ottawa*. Accessed February 8, 2018, from http://www.who.int/healthpromotion/conferences/previous/ottawa/en/

WHO, International Development Law Organization, O'Neill Institute for National and Global Health Law at Georgetown University, The University of Sydney. (2017). *Advancing the right to health: The vital role of law*. Geneva: World Health Organization. Accessed February 8, 2018, from http://www.who.int/healthsystems/topics/health-law/health_law-report/en

WHO-UNICEF. (1978). *Primary health-care. Report of the International Conference on Primary Health Care*. Alma-Ata, USSR. Geneva: World Health Organization. http://www.who.int/publications/almaata_declaration_en.pdf

Wiist, W. (2011). The corporate playbook, health, and democracy: The snack food and beverage industry industry's tactics in context. In D. Stuckler & K. Siegel (Eds.), *Sick societies: Responding to the global challenge of chronic disease* (pp. 204–216). Oxford: Oxford University Press.

What Is and What Is Not a Conflict of Interest in the Sphere of Public Health

Miguel Ángel Royo-Bordonada and Fernando García-López

Abstract A conflict of interest is a set of circumstances or relationships that create the risk that professional judgment or actions regarding a primary interest will be unduly influenced by a secondary interest. Secondary interests may be internal, meaning limited to the personal sphere, or external, linked to the interests of third parties or institutions. Internal interests are related to individual identity and personality, and tend to be perfectly aligned with the primary professional interest. Therefore, we usually only speak of a conflict of interest when external secondary interests, usually of a financial nature, are opposed to the primary professional interest (preserving the health of the population in the case of public health). In addition, biases due to internal interests are bidirectional and difficult to quantify, while those deriving from financial conflicts of interest are predictable, unidirectional and large-scale. Although they are different phenomena, they are often mixed up with one another to create confusion and divert attention from financial conflicts of interest. In public health, conflicts of interest can arise in the field of research when studies are conducted on the health effects of pharmaceuticals, chemicals, tobacco, alcohol or food and in decision-making regarding policies related to these same products. In this chapter we present several examples of conflicts of interest in public health and a series of recommendations for their management, beyond their public declaration, which is necessary but insufficient. All public health organizations should adopt standards to avoid, detect, eliminate or minimize the impact of conflicts of interest and ensure that these are put into practice.

Keywords Conflict of interest · Ethics · Public health · Public health policies · Research

---

M. Á. Royo-Bordonada (✉)
Spanish National School of Public Health, Institute of Health Carlos III, Madrid, Spain
e-mail: mroyo@isciii.es

F. García-López
National Centre of Epidemiology, Institute of Health Carlos III, Madrid, Spain
e-mail: fjgarcial@isciii.es

J. Vallverdú et al. (eds.), *Philosophical and Methodological Debates in Public Health*, https://doi.org/10.1007/978-3-030-28626-2_14

Although the term "conflict of interest" has been interpreted in different ways in the sphere of health, and not always consistently, the definition proposed by the Institute of Medicine (IOM) in the United States reflects the essence of the concept and has been used as a reference in the drafting of numerous institutional policies designed to prevent or manage conflicts of interest among health professionals (IOM 2009). According to this definition, which is widely accepted in the academic world, a conflict of interest is understood as "a set of circumstances that creates a risk that professional judgment or actions regarding a primary interest will be unduly influenced by a secondary interest". What is meant by "undue influence" is a matter of judgment, and not quantifiable; its assessment depends on the context and must be carried out following explicit criteria and using clear arguments. Therefore, as the IOM itself points out, a conflict of interest has the following three components:

1. A primary interest.
2. A secondary interest.
3. The conflict itself.

The mainstream of the scientific literature distinguishes between external (i.e. those deriving from financial interests) and internal (i.e. those deriving from a professional's interest to advance their career) conflicts of interest (Stead 2017). This is an appropriate division for distinguishing between "interests" of internal origin (linked to our beliefs, opinions, personality traits, visions, approaches and expectations of life, which form a more or less permanent part of a person's character and have conditioned the choice of a certain profession) and those of external origin (originating from relationships with third parties—people or organizations—which may or may not have some permanence over the course of our professional life). Ultimately, all personal interests would be, to a greater or lesser extent, of external origin, since they arise from interaction with our family, educational, community and social environment during our childhood, the period when our personality is growing and developing, unless one exists that is fully genetically predetermined. However, for pragmatic purposes, which is what interest us here, the division between internal and external—outside—interests would seem to be a very useful one. That said, and contrary to general opinion, our thesis is that this type of distinction is not applicable to conflicts of interest, because they are always of external origin. In the next section, following a path opened by just a few authors (Bero and Grundy 2016), we present the arguments that support this theory, namely, that internal interests are closely interwoven with our primary professional interest, and so-called internal conflicts of interest are therefore a fiction invented to create confusion and divert attention from financial conflicts of interest.

The Primary Interest in Public Health

Public health is the science and art of preventing disease, prolonging life, and promoting health through the organised efforts of society (Acheson 1988). Public health has a series of distinctive features that are of great relevance from the perspective that concerns us here (Royo-Bordonada and Román-Maestre 2015):

1. Public health interventions are not aimed at individuals, but groups, such as communities and entire populations (education campaigns, vaccinations, screenings). Therefore, an inappropriate decision might have a great impact on public health, since it will affect a lot of people.
2. Public health always involves an action, often governmental, aimed at preserving the health of the population. Therefore, one of the responsibilities of public health officials is to develop health policies and advise governments in carrying out this task.
3. Public health entails collective action by the organized community. Therefore, it has a social component that demands transparency and citizen participation in preparing recommendations for the population, designing interventions and drafting policies.

Given that the primary interest of public health is to preserve and promote the health of the population, mainly through governmental actions, the process of drafting and implementing public health policies should be carefully protected from possible undue influences related to vested interests.

The Secondary Interest

Secondary interests can be economic, professional, family or community-oriented, religious, political or philosophical. Natural and inherent to all people, they move us to establish and pursue our life goals. These interests are related to—or shaped by—our personal experiences: the type of education received, the family model we have been raised in, the religious beliefs that have been passed on to us, the friendships we have made, the community references we have had (teachers, local politicians, community leaders, priests, etc.), the prevailing political ideas and the paradigms of academic, scientific and philosophical thought dominant in our society and in our professional environment. The result is the development of personality and individual identity, with one's own interests, determined by particular preferences and personal convictions.

When defending the thesis that internal conflicts of interest do not exist, it is essential to distinguish between internal and external secondary interests. When a secondary interest is confined to the personal sphere, we speak of internal interests; when it is mediated by or linked to the interests of third parties—people or institutions—, we speak of external interests. For example, the pecuniary interest can be internal (the expectation of receiving a salary for our work), or external, when we receive additional financial compensation for our work from someone outside the organization that employs us. Religious interests can also be internal, when they are circumscribed to the intimate sphere of personal beliefs, or external, when one is an active member of a religious organization. And the same goes for community or political interests. The key difference between the two is the feeling of obligation towards third parties, whether due to a demand deriving from the type of relationship established, as may occur in religious congregations or political parties, or by a sense of gratitude or reciprocity, which may even be induced unconsciously through small gifts (Dana and Loewenstein 2003).

The Conflict

According to the IOM definition, the conflict itself "is the set of circumstances or relationships that create or increase the risk that the primary interest will be neglected as a result of the pursuit of secondary interests" (IOM 2009). Although it may seem obvious, it is worth pointing out that the *sine qua non* requirement for a conflict of interest to exist is that the two interests at stake are opposed in some way, that is, the pursuit of one results in a weakening of the other. Some authors have even proposed the term confluence rather than conflict of interest (Cappola and FitzGerald 2015), which would be inappropriate, since the conflict occurs precisely when there is no confluence of interests. This only contributes to further increase the confusion existing around the issue (Shaw et al. 2016).

The existence of conflict through family interests is obvious when one is in charge of recruitment for public positions or the allocation of public resources. Although there may well be families not prone to nepotism or even very critical of this type of behaviour among their members, to the point of reaching the opposite extreme, feeling gratitude and reciprocity towards people from our inner circle cannot be avoided. Hence the existence of legal provisions to prevent this type of undue influence through close personal relationships, such as those with relatives, friends, colleagues and rivals (Howard 1993). The conflict of interest can also occur through relationships with organizations or institutions, as in the case of a doctor belonging to a religious congregation that demands its members try to dissuade the population in their environment, including in their professional practice, from the use of condoms as a means of preventing sexually transmitted diseases (STDs) or unwanted pregnancies.

It is worth making a qualification with regard to financial interests, which are the main source of conflicts of interest (Bero and Grundy 2016) in research and professional practice. On the one hand, there is the natural (internal) interest in receiving a fair wage in exchange for our work. In theory, what we expect is that the interest of our employer (whether the public administration, a private company, a foundation or an association) fully coincides with our primary interest and that is why they pay us to do our job well. Also, the better professionals we are, the more valued we will be and the more they will pay us for our work. Therefore, in the absence of competing interests, there is no room for conflict of interest. However, there may be partial exceptions to this principle, such as capitation or payment for service systems, with incentives to apply more or less care than required (Larkin and Loewenstein 2017), or corporate capture phenomena with public administrations, whereby private interests prevail over those of the population in the drafting of health policies (Mindell et al. 2012). Furthermore, there are also outside financial interests, when people or organizations other than our main employer are interested in the results of our professional activity and pay us an additional fee for doing our work, either through direct financial compensation or gifts, or by financing teaching or research activities (Dana and Loewenstein 2003). These external financial interests are often in obvious opposition to the primary interest of public health, as in the following examples:

- Tobacco or soft drink companies funding researchers who are studying the effects of passive smoking or sugar-sweetened beverages consumption on health (Brownell and Warner 2009).
- Spirits companies financing official Ministry of Health campaigns for the prevention of alcohol consumption, as recently occurred in Spain (Royo-Bordonada 2014).

In scenarios such as these, no one doubts that the professionals who sign up for such collaborations, whether researchers or public health officials, are subject to a conflict of interest, and consequently recommendations are in place to not accept funding from sources whose primary interests run counter to public health interests (Galea and Saitz 2017). However, competing interests are not so obvious in relations with the pharmaceutical industry due to the existence of shared interests, such as trying to improve patients' adherence to therapy, so the context must always be analysed to assess possible hidden conflicts of interest. One example would be the 2013 American guidelines for assessing cardiovascular risk (The American College of Cardiology/American Heart Association), which recommend taking statins to reduce cholesterol levels as a primary prevention when the estimated risk of cardiovascular disease exceeds a certain percentage—more than 7.5% estimated over a period of 10 years. Of the 15 members of the committee that wrote the guidelines, eight, including the president and two co-chairs, had links to pharmaceutical companies that sold medications to lower blood cholesterol (Lenzer 2013). Some of these members broke ties with the industry during the writing of the guidelines and agreed not to establish new links for 2 years after their publication. The members who had connections with the industry did not disclose them at the time the committee was formed; rather, these links were revealed as the result of an investigation by a medical journal. These guidelines propose the mass use of statins as primary prevention of cardiovascular disease in billions of people around the world (Ioannidis 2014). The presence of these links with the industry may have exaggerated the role of pharmacological measures to the detriment of other non-pharmacological measures in cardiovascular prevention. It is clear that the authors of the guidelines should have no links with pharmaceutical companies interested in their content. In these cases, we can suspect the existence of a conflict of interest when a reasonable person would interpret that the circumstances are potentially sufficient to influence the judgment of the professional in question (Fineberg 2017); for example, where the pharmaceutical industry provides purpose-oriented funding for research, scientific conferences, informative activities or training courses.

In the previous paragraphs, we have seen several examples where outside interests, particularly financial ones, clashed with the primary interest, causing a conflict of interest. In contrast, the desire to progress in one's public health professional career and improve one's personal reputation, which is often cited as a prime example of a source of conflict of internal interests (Stead 2017), does not conflict with the primary interest of improving the health of the population. Quite the contrary, in fact: the more qualified professionals are and the better their research, the greater their prestige and the greater the likelihood they will progress. That is to say, in this case there are no competing interests, and consequently there is no conflict of interest whatsoever. A very different scenario would be a researcher falsifying or inventing results to progress professionally in a fraudulent manner. But that would be a lack of scientific integrity, a different phenomenon, which is not comparable to conflicts of interest.

Unlike the desire to advance in one's professional career, there are other internal interests that, viewed from the ideal perspective of an impartial observer, may collide or not be perfectly aligned with the primary interest of a public health professional, such as religious beliefs, political ideas or support for certain scientific theories or philosophical or moral views on health and well-being. An extreme example of this would be a personal moral belief that rejects the use of condoms to prevent STDs and unwanted pregnancies or defends gay conversion therapies, despite their absence of benefits and harmful effects (American Psychological Association 2009). This resembles the aforementioned case of scientific fraud, where even if there are competing interests we do not speak of conflicts of interest, but rather of a lack of professional integrity, since personal beliefs or convictions are placed before professional duties, causing damage by omission (relinquishing the obligation to recommend condom use) or by action (recommending conversion therapy). What would constitute a case of conflict of interest would be an honest professional who works according to purely scientific criteria being a member of a religious congregation that pressures him to act in the opposing direction (outside interest) to that of his professional ethics.

Adhering to a political ideology and a certain school of thought influence the way we approach professional practice. In the field of research, scientific hypotheses are guided by theories assimilated from our training and previous life experiences, particularly in a field such as public health, which is closely linked to the social sciences. To state that by sustaining a certain hypothesis one is subject to a conflict of interest would be akin to stating that all scientific research is subject to conflicts of interest or that all individuals have a permanent conflict of interest with themselves. Internal interests are an essential, inherent and inseparable part of our individual identity, that backpack that we carry for life with everything that allows us to advance and progress as human beings. One might say, as some claim, that at the individual level there is no absolute scientific objectivity but a reasonable objectivity, since, except in an imaginary world, nobody can approach research hypotheses in a completely impartial manner, detached from their own identity and their defence of certain theories they consider more convincing than others (Vineis and Saracci 2015). In fact, there is nothing more partial than the pretence of impartiality, as illustrated by the Spanish proverb cited by Schopenhauer, "A clattering horseshoe lacks a nail", which refers to the person who "boasts much of his nobility, science, virtue, and has considerable faults on it" (Schopenhauer 2014). Absolute objectivity in the scientific world can only be achieved at the collective level, not as an individual enterprise. In fact, it is the individual interests of researchers to generate unbiased knowledge from necessarily partial hypotheses that allows science to advance as a whole. And this is so because the essence of science is to combat the natural tendency towards a partial vision of the world through rigorous research; therefore, good researchers, aware of their own biases, decide to test the hypotheses deriving from their theories in an attempt to refute them, and they are willing to accept others hypothesis more consistent with the empirical results, even if they go against their previous theories. Therefore, internal interests can also produce biases, and good examples of this are the well-known biases of self-service, confirmation and publication (Rosenbaum 2015a; Raymond 1998; Dwan et al. 2013). But these biases can be avoided, or at least minimized, with good doses of professional integrity, which entails the willingness to apply the scientific method to its ultimate consequences and a good knowledge of tools for preventing the introduction of biases (George et al. 2016). In the words of Popper, "Those among us who are unwilling to expose their ideas to the hazard of refutation do not take part in the scientific game (Popper 1959). However, the sociology of scientific research indicates that researchers are immersed in patterns generated by the paradigms in which they work while often being blind to outside paradigms, which coexist without the former accepting them (Kuhn 1970). This is so to the point that old paradigms sometimes die together with the natural death of their defenders. Therefore, scientific integrity allows us to minimize the risk of bias and accept the refutation of our hypothesis and its replacement by new ones, albeit usually within the framework of the same scientific paradigm.

When one attempts to play without accepting the rigorous application of the rules of the scientific method, whether for religious, political, philosophical or moral reasons, then we are dealing with a lack of scientific integrity, which is quite different from a conflict of interest, even if they are sometimes confused as if they were different degrees or levels of the same phenomenon (Stead 2017). Similarly, when we put our beliefs, ideas or personal convictions before our professional obligation to ensure public health, we are dealing with a case of malpractice. Rosenbaum's series of articles on conflicts of interest published in one of the most widely circulated medical journals (Rosenbaum 2015a, b, c), and with the endorsement of the journal's director (Drazen 2015), constitute an attempt to minimize the importance of conflicts of interest between the pharmaceutical industry and the fields of medicine and biomedical research by diluting them in a jumbled up mess of biases and problems related to scientific integrity. Conflicts of interest happen regardless of whether a particular individual or institution is consciously influenced by a secondary interest. Therefore, this does not mean that most doctors or researchers are influenced by the

Table 1 Possible presence of bias depending on the level of integrity and existence of conflicts of interest

		Conflict of interest	
		YES	NO
Level of integrity	High	++	+
	Low	++++	+++

economic gains they can obtain, only that it is often difficult, if not impossible, to distinguish between those cases in which economic gains have undue influence and those in which they do not (Thompson 1993; Steinbrook et al. 2015). It is surprising that the journal adopted such a clear position when it was in fact the first one to raise the importance of conflicts of interest (Relman 1980) and to propose measures for managing them (Relman 1984). In this sense, although the proper handling of conflicts of interest is a key element in ensuring the scientific integrity of researchers, their nature differs from the usual cases of malpractice, such as fabrication, plagiarism or falsification. Although in the presence of a conflict of interest there may be bad faith and lack of scientific integrity on the part of the interested parties if they consciously allow themselves to be corrupted for the economic benefits they may obtain, in many cases conflicts of interest affect the objectivity of researchers' judgment in an unconscious way. This potential unconscious influence is what distinguishes conflicts of interest from classic problems of scientific integrity. Finally, conscientious objection, when included in a regulation, covers omission of duty (provided that the action being objected to is carried out by third parties) but not malpractice by action. In summary, internal interests and the existence of conflicts of interest or the lack of professional integrity are all associated with a potential risk of bias, but in different degrees, since biases due to scientific fraud are in no way related, either quantitatively or qualitatively, to unintentional, and more or less unconscious, biases deriving from a conflict of interest and even less to an unconscious bias deriving from an internal interest (Table 1).

Potential Biases Due to Internal Interests Are Quite Different from Those Due to Conflict of—External—Interest

Following the reasoning in the previous section, it could be assumed that, in the field of research, as happens with potential biases deriving from internal interests, a willingness to apply the scientific method would be sufficient to avoid or minimize the risk of bias deriving from conflicts of interest. However, all other conditions remaining equal, including scientific rigour measured using the instruments available for analysing the internal validity of studies, the results of studies with financial conflicts of interest are systematically more favourable to the interest of the industries involved than those of studies not subject to conflict (Barnes and Bero 1998). Why, then, does rigorous application of the scientific method not work in the case of conflicts of interest? Why is so difficult to prevent biases due to conflicts of interest? The answer is simple: With internal interests one is alone with oneself, and having a good knowledge of the scientific method, including common tools for preventing biases, and the willingness to apply it in a rigorous manner (integrity) are enough to manage the situation and minimize the risk of bias; on the contrary, in the presence of external interests, a person of equal integrity (with the same willingness to rigorously apply the scientific method) is at the same time subject to an obligation or a sense of gratitude or reciprocity towards an external agent. This can condition, in a more or less unconscious way, the introduction of biases that are less obvious at first glance and undetectable with the methods used to evaluate the internal validity of studies, such as those that can arise in the hypothesis formulation (e.g. if that particular hypothesis had not been considered or had been considered differently in the absence of conflict of interest) or due to subtleties in the design of the study, the defining of selection criteria for participants, the selection of methods of analysis and outcome variables, and the presentation of the results. In addition, internal interests can be adapted according to new experiences, in the same way that a scientific hypothesis is replaced by another one or modified to adapt to new scientific evidence; while outside interests, financial or otherwise, are not modified according to the experiences or results of an investigation, since they are primary interests—alien or opposed to the health of the population—of third parties.

In order to illustrate this last point, we will use the example of the study of a possible causal association between the consumption of sugary drinks and obesity. In 2010, the prestigious *International Journal of Obesity* published an article that presented a type of bias denominated "white hat bias" by its authors (Cope and Allison 2010), who defined it as "a bias leading to the distortion of scientific information in the service of what may be perceived to be righteous ends (righteous meaning morally good and correct)". The authors not only sought to highlight the existence of a possible bias deriving from an internal interest, but also tried to quantify it. This was not a new type of bias, but rather a combination of confirmation and publication biases. To illustrate the existence of this bias in research on the relationship between sugary drinks and obesity, Cope and Allison conducted three types of analysis:

1. Analysis of citations. After selecting two original articles on the subject, published in 2004 and 2006, both with results that were significant for one indicator of obesity (body mass index) and not significant for another (i.e. body fat percentage), Cope and Allison proceeded to review the subsequent articles that cited them, concluding that the vast majority exaggerated the beneficial effect of reducing the consumption of sugary drinks when describing the results of the cited article. Although these results are very suggestive of confirmation bias, the analysis suffers from two important limitations. In the first place, a systematic review was not carried out to exhaustively searching for all the articles of interest and the description of the bibliographic search method was clearly insufficient, being limited to the following sentence: "A search on the *Web of Science* (in October 2008) found 195 articles citing James et al. and 45 citing Ebbeling et al., respectively". And secondly, there is no evidence of the validity and reliability of the instrument used to analyse the citations, a fundamental aspect when analysing a response variable with a strong subjective component, such as

the degree to which the description of the results of a study is balanced when quoted in an article (Kulkarni et al. 2009).

2. Analysis of publication bias. This analysis was conducted using data from a meta-analysis published in 2007 (Vartanian et al. 2007). In the words of Cope and Allison: "the results suggest the presence of a publication bias, which seems to only be present in studies not funded by the industry". Although the original systematic review included cross-sectional, longitudinal, and intervention studies, the analysis of publication bias was limited to the less informative cross-sectional ones, without explaining the reasons for this arbitrary selection. This decision is especially worthy of note if we take into account that the effects observed in the meta-analysis were greater in the longitudinal and experimental studies, as the authors highlight in the summary and discussion of their article. In addition, the authors of the systematic review had conducted their own analysis of a potential publication bias, and suggested there was only a low probability of it. It is worth noting that none of these fundamental aspects of the results of the systematic review were mentioned in the article by Cope and Allison, which would thus suffer from the same problem of inappropriate citation that they had previously tried to disclose in other articles to demonstrate the existence of white hat bias.
3. Analysis of inappropriate communications in press releases. This analysis is limited to reporting a limited number of supposedly inadequate, incomplete or exaggerated communications. Although the references presented suggest confirmation bias, the fact that they are anecdotal cases analysed without any scientific methodology means they are difficult to assess.

In recent years, several systematic reviews and meta-analyses have been carried out in an attempt to elucidate the effect of sugary drinks on the risk of obesity, with diverging results. Although the quality of the reviews does not seem to be affected by the study's source of funding (Massougbodji et al. 2014), the likelihood of concluding that sugary drinks do not cause obesity can be up to five times higher in industry-funded reviews (Bes-Rastrollo et al. 2013). Studies funded by the pharmaceutical industry also show more favourable results towards their products than independent studies, which cannot be explained by methodological differences, at least not with the usual tools for evaluating biases (Bero 2013). This apparent paradox can be explained by the existence of more subtle biases, such as the inclusion and exclusion criteria used in the studies (excluding those cases that could respond worse to the intervention studied), the comparator used in the control group (use of an inferior comparator or in insufficient doses), the type of statistical analysis chosen (not strictly carrying out an intention-to-treat analysis) and the way in which the results were presented (selective description of only those results that favour the commercial interests of the funding organization). Similarly, discrepancies in reviews on the effect of sugary drinks can be explained by the decision to include original studies according to the type of analysis performed, since those studies that adjust for energy intake tend to find no or smaller effects (Trumbo and Rivers 2014; Crowe-White et al. 2016). However, adjusting for energy is not appropriate, since it is a mediator of the association between the intake of sugary drinks and obesity (Malik et al. 2009). In fact, the hypothesis of the possible causal effect of sugary drinks is based on previous data related to the low satiating capacity of liquid calories, which are not compensated by a reduced intake of other foods, and the consequent increase in total caloric intake associated with their consumption. Currently, the weight of scientific evidence clearly leans towards the existence of a causal relationship between the consumption of sugary drinks and obesity (Keller and Bucher Della Torre 2015; Ruanpeng et al. 2017). Therefore, the most plausible hypothesis is that the lack of initial agreement between different researchers did not derive from a white hat bias, but from a bias due to financial conflict of interest. This should not surprise us, if we consider that shortly after presenting the white hat bias it was made public that Allison had received more than $2 million from soft drinks and fast food companies to fund his research studies (Harris and Patrick 2011).

Based on all of the above, we can infer that biases deriving from conflicts of interest, and particularly financial ones, which are also the most widespread, present the following distinctive elements (Bero and Grundy 2016):

1. They even occur with small gifts, which are enough to create a feeling of obligation and reciprocity that affects researchers' judgment, often unintentionally and unconsciously.
2. They are predictable, since they always go in the same direction, in favour of the secondary interest, that of the funding industry.
3. They are objective and quantifiable. A host of studies have repeatedly and consistently demonstrated the existence of such kind of biases, which have been precisely quantified in relation to the interests of the pharmaceutical, chemical, alcohol, tobacco, and food industries.
4. They are of overwhelming size, incomparable in any way with possible biases from internal interests, whose influence is limited to the sphere of the individual. Therefore, their impact can be very large, since they have the potential to bias all scientific results in a particular field of knowledge.
5. They can be controlled, by eliminating the conflict of interest and renouncing the funding in financial conflicts, or through recusal processes in personal and institutional conflicts.

The Ceremony of Confusion

All people have internal interests, and in a plural and democratic society it is to be hoped that these interests will go in different (unforeseeable) directions and the potential bias deriving from them, such as confirmation and publication biases, may, at least in part, compensate among themselves, since there is evidence that such biases are not unidirectional (Ditto et al. 2017) and that left and right-wing people both tend to avoid exposing themselves to the opinions of those who have political ideas different from their own (Frimer et al. 2017). In addition, researchers' internal interests are essential to the advancement of scientific knowledge, acting as a breeding ground for scientific hypotheses that, once contrasted, will allow new refined hypotheses to be generated, and so on. If this phenomenon has nothing to do with conflicts of interest, either quantitatively or qualitatively, and particularly with those of a financial nature, it seems logical that they be analysed separately. So why do we often find the two mixed up in the scientific literature?

The main objective of this self-serving ceremony of confusion is to divert attention from conflicts of financial interests and dilute the problems caused by this phenomenon in the mud waters of possible biases due to internal interests. For example, more than half of the major clinical journals now demand the declaration of at least one non-financial conflict of interest (Shawwa et al. 2016). This is all about playing down the importance of financial conflicts of interests (Serra Majem 2016) by treating them as just another among equals and consequently requesting a comprehensive regulation of financial and non-financial conflicts of interest (Saver 2012), ignoring the fact that biases due to financial conflicts of interest have been quantified in a great variety of contexts, always go in the same direction and have a great capacity for dissemination, with amplifying effects that can dominate a specific field of research (Bero and Grundy 2016). Secondly, sowing doubt about all researchers erodes population confidence in science. And thirdly, if any interest is catalogued as a conflict of interest and all people are subject to them, the phenomenon is presented as immeasurable, completely widespread and therefore unmanageable. There are even those who go further and argue that professionals with external financial interests have the magical capacity to manage them effectively precisely because they are subject to them (Serra Majem 2016). This goes against all logic and the results of the scientific studies mentioned previously. The next step is to cast suspicion on the integrity of researchers who do not have external financial interests, through *ad hominem* judgements alluding to alleged "internal conflicts" of an obscure nature, as if those who accused them did not have internal interests as all other mortals do (Serra Majem 2016). As absurd as it may seem, this strategy is bearing fruit. A good example is what are known as intellectual conflicts of interest. This concept, badly conceived and defined, allows an expert to be accused of a lack of impartiality for having a particular and strong point of view on a matter in question, precisely the opposite of that of their accusers. This finding is the result of an analysis of recusals due to intellectual conflicts of interest on the Food and Drug Administration (FDA) advisory committees, where total asymmetry was found in their application, always in favour of industry interests (Lenzer 2016). While numerous experts have been excluded from the committees for questioning the safety or efficacy of the products analysed, no cases have been documented of an expert being recused for defending the efficacy of a new drug or medical technology.

Types of Conflicts of Interest in Public Health

Conflicts of interest in public health mainly arise in two areas. First, in research, when we study the health effects of products marketed by private businesses, such as pharmaceutical, chemical, tobacco, alcohol or food companies. Second, and more importantly, because it can affect millions of people, in the development of policies, especially those related to the aforementioned products (Gupta et al. 2015). Among possible sources of conflicts of interest in the development of public health policies, the following have been described: industry experts participating on committees that establish guidelines and recommendations or regulatory bodies, the revolving door phenomenon or the ownership of shares in companies by those responsible for their regulation and, last but not least given the magnitude they are acquiring and the way in which they are infiltrating health systems, public–private partnerships in the design and implementation of public health policies (Hernandez-Aguado and Zaragoza 2016).

In a previous section of the chapter, we provided a detailed description of biases in studies into the relationship between sugary beverages and obesity caused by a conflict of interest in the field of research. As a sample of conflicts of interest in the drafting and implementing of public health policies, below we will describe the action taken by the World Health Organization with regard to the A H1N1 2009 influenza pandemic and the way in which the Spanish Agency for Food Consumption, Safety and Nutrition (AECOSAN) addressed the national strategy for nutrition, physical activity and prevention of obesity (the NAOS Strategy).

A H1N1 2009 Influenza Pandemic

When the first reports of cases came in from Mexico, the idea spread that the flu pandemic could be extremely lethal and perhaps reach the dimensions of other pandemics that had caused mass mortality, such as the so-called "Spanish flu" of 1918. However, expectations regarding the appearance of a lethal pandemic dated back to 10 years previously. During that period it was thought that the question was not whether there would be a pandemic but when one would occur. And a lot of money was spent from public budgets in preparation for the anticipated pandemic, which it was expected would be very serious (Doshi 2009). On June 11, 2009, the WHO raised the pandemic alert level to phase 6, the highest possible. In making its pandemic declaration, the WHO recommended cases be treated with antiviral oseltamivir (Tamiflu) and zanamivir (Relenza) and the mass vaccination of the population. Many countries bought huge amounts of antivirals and vaccines with their public healthcare budgets. In the end, the pandemic turned out to be very benign and many of the drugs bought and stored were not put to use.

Subsequent research revealed that many of the experts who advised the WHO had undeclared conflicts of interest with the antiviral and vaccine manufacturers (Cohen and Carter 2010; Godlee 2010). It was also discovered that WHO Director General Dr. Margaret Chan received advice from a committee whose members were kept secret even for other WHO advisers who participated in the process. Also, during the WHO's preparations in the previous 10 years it had worked closely with the European Scientific Working Group on Influenza (ESWI), a group of experts that was assembled to "combat the impact of the flu epidemic and pandemic" and was 100% financed by the pharmaceutical industry, among others by companies that supply antivirals and act as a pressure group in their favour. Finally, the supposed effectiveness of the antivirals was not as high as expected, because very little evidence was found in its favour at the time. Subsequently, a Cochrane systematic review of oseltamivir only found a slight reduction in the duration of the symptoms and no reduction in the frequency of complications (Jefferson et al. 2014).

From the viewpoint of anyone judging the events a posteriori, the WHO's predictions were very exaggerated. It estimated an affectation of two billion people worldwide, but the A H1N1 2009 influenza pandemic hardly resembled that expected, as the virus subtype had been circulating for many years and a good part of the population was already immune to it. In addition, although the outbreak had been intense in Mexico, albeit with doubts over the veracity of the data, it was quite mild in the first affected countries of the southern hemisphere, with little virulence or infectivity. However, the truth is that there was a worldwide sense of generalized panic in a context of enormous uncertainty. When conflicts of interest were discovered within the WHO, the Council of Europe criticized its lack of transparency and drew attention to a series of negative consequences of its actions, such as the distortion of public health priorities throughout Europe, the waste of huge amounts of money in the purchasing of unused pharmaceutical products, the spread of unfounded fear and the exposure of many people to the potential risks of vaccines that had not yet been adequately tested (Flynn 2010). And a further very negative consequence of all this was the population's loss of confidence in the health authorities.

In summary, people with conflicts of interest with pharmaceutical companies that had interests in the flu participated in the decisions taken by the WHO regarding the treatment and prevention of flu without the population or governments knowing the existence of those conflicts at the time. The implementation of the WHO's recommendations had a series of serious negative consequences in the field of health, the economy and the credibility of the public health authorities.

The NAOS Strategy

In 2005, the Spanish Agency for Food Safety, the predecessor of AECOSAN, launched its NAOS Strategy, in which the food and advertising industries, among others, played a significant role (Spanish Agency for Food Safety 2005). One of the declared aims of the strategy was to promote public–private partnerships (Neira and de Onis 2005). In the words of the then director of the Agency and current WHO public health director, "The promotion of a healthy diet and an active life-style must not be tied to a repressive campaign, interpreted by citizens as being prescriptive and based on restrictions. It is not about good or bad food, but rather about a well- or poorly-balanced diet" (Neira and de Onis 2006). That is to say, the strategy invoked the well-known food industry mantra that there are no good or bad foods to suggest that, contrary to scientific opinion, any foods can form part of a balanced diet, and pointed to the direction future nutrition and obesity-prevention policies were to follow based on industry self-regulation and educational measures. The code for the self-regulation of food advertising aimed at minors (known as the PAOS code, for its acronym in Spanish) was the first measure to be applied, paving the way for other measures based on voluntary recommendations that would follow. Subsequently, the former director of scientific and normative issues for Coca-Cola Iberia was appointed executive director of the Spanish Agency for Food Safety and Nutrition, giving rise to a conflict of interest that was repeatedly reported without the government taking any action, except to appeal over the legality of the appointment and to keep her in her position (Royo-Bordonada 2014). Ten years after the launch of the NAOS strategy, its policies based on public–private partnerships with the same industries that profit from the sale of unhealthy foods and beverages, whose consumption contributes to the epidemic of obesity and non-communicable diseases, have proven to be ineffective (Royo-Bordonada 2016). At the same time, the World Health Organization, endorsing experts' recommendations, has launched several appeals to Member States to implement effective policies, such as taxes on sugar beverages, subsidies for healthy food, restrictions on food advertising, interpretive front of pack labelling and prohibition of the sale of unhealthy products in schools, sports centres and hospitals (WHO 2016). However, today, nutrition policies in Spain continue to focus on voluntary industry self-regulation and education measures, which place emphasis on individual responsibility, bypassing social responsibility in the way we feed ourselves (Table 2). And this is true even though social responsibility is much more broad and far-reaching than individual responsibility, given the numerous and varied range of environmental factors beyond the control of the individual that condition our eating habits. Of these, the following are worth mentioning: the production and distribution of a wide range of cheap and ultra-processed products with nutrients and flavour enhancers to amplify the reward mechanism associated with their intake; their intensive marketing by means of deceptive advertising techniques; large ration sizes; the use of complex nutrition labelling which is difficult to understand for the average citizen, with health claims and scientific or professional endorsements; and the transformation of schools, hospitals and sports centres into commercial centres and distributors of these products through the generalized installation of vending machines (Royo-Bordonada 2017). In summary, the divergence between the WHO's policy proposals for nutrition and the measures applied by the Spanish Government comes down to conflicts of interest for public health officials, managers and politicians of AECOSAN, who allow the private, purely economic interests of the food and advertising industries to exert undue influence on the development and implementation of nutrition policies, which should be guided by public health criteria alone.

Table 2 Nutrition policies proposed by the World Health Organization and the current state of affairs in Spain, 2018

World Health Organization proposals	Spain
Implement an effective tax on sugary beverages	Only applied in Catalonia
Implement WHO recommendations to restrict advertising of unhealthy foods and beverages aimed at the child population	The PAOS Code does not regulate the nutritional profile of advertised products
Develop nutritional profiles to identify unhealthy foods and beverages	Not developed
Implementation of interpretive front-of-pack labelling systems, like nutritional traffic lights	Not implemented
Prohibit the sale of unhealthy foods and beverages at schools, day-care centres, sports centres and children's events	Partial and variable implementation in some autonomous regions
Increase disadvantaged communities' access to healthy foods	There are no comprehensive initiatives that address barriers to access

Source: Adapted from Royo-Bordonada MÁ. Challenges of nutrition policies in Spain: from individual responsibility to collective responsibility. Rev Esp Nutr Hum Diet. 2017;21(Supl. 1):56-8

Management of Conflicts of Interest

The handling of conflicts of interest is usually limited to their simple declaration. However, mere declaration is insufficient, proof of this being that those who acknowledge the existence of the conflict tend to believe they are affected by them to a lesser extent than their colleagues, which is contrary to empirical data (Steinman et al. 2001; González-Rubio et al. 2017). If they consider themselves immune to the conflict to a certain degree, they cannot be expected to effectively manage the inevitable biases associated with it. Although acknowledging the problem is a first step, it does not resolve it because it does not serve to regulate it, and, in the absence of another type of regulation, it can have negative perverse effects, such as bias in exaggerating the results to compensate for the mistrust that may have been created by the declaration of a conflict of interest, or the effect of moral permissiveness that they may have in the scientific community, because once they are declared, there is a risk of the assumption that anything goes, as if the declaration actually meant the removal of the conflict (Loewenstein et al. 2012). The public declaration of conflicts of interest is a necessary but insufficient condition. All organizations (research institutions, funding agencies, scientific journals, peer review boards) should have rules to avoid, detect, eliminate and minimize the impact of conflicts of interest, and ensure that these are put into practice (Bero and Grundy 2016). Among the possible measures that these rules could incorporate, the following have been mentioned:

1. Do not accept funding or collaborate with companies with interests that are opposed to those of public health.
2. When there is no apparent opposed interest, independent observers should analyse the possible existence of hidden competing interests.

3. Do not accept purpose-oriented funding. Private funding of research, training or health promotion activities should go to a common fund, managed by an independent body.
4. Prevent people who have previously worked in private companies from occupying positions of responsibility in drafting recommendations and decision-making that affect their products (the revolving door phenomenon).
5. Prevent anyone with conflicts of interest from active participation in the development of consensus statements, clinical practice guidelines, statements of principles or technical recommendations for health authorities.
6. Produce and disseminate rules on the prevention and management of conflicts of interest in all educational, research and administration institutions related to public health.
7. Incorporate rules for the prevention and management of conflicts of interest into research institutions' codes of good scientific practices.

In addition to declaring and establishing rules with the aforementioned measures, good scientific and ethical training is essential for the prevention, detection and management of conflicts of interest, with particular emphasis on anything related to conflicts of interest and their impact on the health of the population. To conclude, then, we invite anyone not fully convinced of the arguments presented here to reflect on the following: if it were possible to talk about the existence of internal conflicts of interest, even if only from a purely conceptual or theoretical point of view, we would be forced to publicly declare our religious beliefs, political ideas and moral convictions, an absurd notion that goes against a fundamental right enshrined in all constitutions in democratic states of law.

Disclaimer This is an independent research, and the opinions expressed are those of the authors and not those of the Institute of Health Carlos III.

References

Acheson, D. (1988). *Public health in England*. London: Department of Health.

Agencia Española de Seguridad Alimentaria. (2005). *Estrategia NAOS. Agencia Española de Seguridad Alimentaria*. Accessed January 12, 2018, from http://www.aecosan.msssi.gob.es/AECOSAN/docs/documentos/nutricion/estrategianaos.pdf

American Psychological Association. (2009). *Task force on appropriate therapeutic responses to sexual orientation*. Report of the American Psychological Association Task Force on Appropriate Therapeutic Responses to Sexual Orientation. Accessed December 20, 2017, from https://www.apa.org/pi/lgbt/resources/therapeutic-response.pdf

Barnes, D. E., & Bero, L. A. (1998). Why review articles on the health effects of passive smoking reach different conclusions. *Journal of the American Medical Association, 279*(19), 1566–1570.

Bero, L. (2013). Industry sponsorship and research outcome: A Cochrane review. *Journal of the American Medical Association Internal Medicine, 173*(7), 580–581.

Bero, L. A., & Grundy, Q. (2016). Why having a (nonfinancial) interest is not a conflict of interest. *PLoS Biology, 14*(12), e2001221.

Bes-Rastrollo, M., Schulze, M. B., Ruiz-Canela, M., & Martínez-González, M. A. (2013). Financial conflicts of interest and reporting bias regarding the association between sugar-sweetened beverages and weight gain: A systematic review of systematic reviews. *PLoS Medicine, 10*(12), e1001578.

Brownell, K., & Warner, K. E. (2009). The perils of ignoring history: Big tobacco played dirty and millions died. How similar is big food? *Milbank Quarterly, 87*, 259–294.

Cappola, A. R., & FitzGerald, G. A. (2015). Confluence, not conflict of interest: Name change necessary. *Journal of the American Medical Association, 314*(17), 1791–1792.

Cohen, D., & Carter, P. (2010). Conflicts of interest. WHO and the pandemic flu "conspiracies". *British Medical Journal, 340*, c2912.

Cope, M. B., & Allison, D. B. (2010). White hat bias: Examples of its presence in obesity research and a call for renewed commitment to faithfulness in research reporting. *International Journal of Obesity (London), 34*(1), 84–88.

Crowe-White, K., O'Neil, C. E., Parrott, J. S., Benson-Davies, S., Droke, E., Gutschall, M., et al. (2016). Impact of 100% fruit juice consumption on diet and weight status of children: An evidence-based review. *Critical Reviews in Food Science and Nutrition, 56*(5), 871–884.

Dana, J., & Loewenstein, G. (2003). A social science perspective on gifts to physicians from industry. *Journal of the American Medical Association, 290*(2), 252–255.

Ditto, P. H., Liu, B. S., Clark, C. J., Wojcik, S. P., Eric, E. C., Grady, E. H., et al. (2017, April 13). *At least bias is bipartisan: A meta-analytic comparison of partisan bias in liberals and conservatives.* Available at SSRN. Accessed January 12, 2018, from https://ssrn.com/abstract=2952510

Doshi, P. (2009). Calibrated response to emerging infections. *British Medical Journal, 339*, b3471.

Drazen, J. M. (2015). Revisiting the commercial-academic interface. *New England Journal of Medicine, 372*(19), 1853–1854.

Dwan, K., Gamble, C., Williamson, P. R., Kirkham, J. J., & Reporting Bias Group. (2013). Systematic review of the empirical evidence of study publication bias and outcome reporting bias – An updated review. *PLoS One, 8*(7), e66844.

Fineberg, H. V. (2017). Conflict of interest: Why does it matter? *Journal of the American Medical Association, 317*(17), 1717–1718.

Flynn, P. (2010). *Social, Health and Family Affairs Committee, Parliamentary Assembly of the Council of Europe. The handling of the H1N1 pandemic: More transparency needed.* AS/Soc, 12. Accessed January 8, 2018, from http://assembly.coe.int/CommitteeDocs/2010/20100329_MemorandumPandemie_E.pdf

Frimer, J. A., Skitka, L. J., & Motyl, M. (2017). Liberals and conservatives are similarly motivated to avoid exposure to one another's opinions. *Journal of Experimental Social Psychology, 72*, 1–12.

Galea, S., & Saitz, R. (2017). Funding, institutional conflicts of interest, and schools of public health: Realities and solutions. *Journal of the American Medical Association, 317*(17), 1735–1736.

George, B. J., Beasley, T. M., Brown, A. W., Dawson, J., Dimova, R., Divers, J., et al. (2016). Common scientific and statistical errors in obesity research. *Obesity (Silver Spring), 24*(4), 781–790.

Godlee, F. (2010). Conflicts of interest and pandemic flu. *British Medical Journal, 340*, c2947.

González-Rubio, R., Escortell-Mayor, E., & Del Cura González I. (2017). *The pharmaceutical industry and specialised medical training: Residents' perceptions in Madrid, Spain*. Gac Sanit. https://doi.org/10.1016/j.gaceta.2017.07.015.

Gupta, A., Holla, R., & Suri, S. (2015). Conflict of interest in public health: Should there be a law to prevent it? *Indian Journal of Medical Ethics, 12*(3), 172–177.

Harris, D., & Patrick, M. (2011, June 21). Is 'Big Food's' big money influencing the science of nutrition? *ABC NEWS*. Accessed January 12 2018, from http://abcnews.go.com/US/big-food-money-accused-influencing-science/story?id=13845186

Hernandez-Aguado, I., & Zaragoza, G. A. (2016). Support of public-private partnerships in health promotion and conflicts of interest. *British Medical Journal Open, 6*(4), e009342.

Howard, F. (1993). Rules and governance in the public sector: The case of civil service. *American Journal of Political Science, 37*(4), 990–1007.

Institute of Medicine. (2009). *Conflict of interest in medical research, education, and practice*. Washington, DC: National Academies Press.

Ioannidis, J. P. (2014). More than a billion people taking statins?: Potential implications of the new cardiovascular guidelines. *Journal of the American Medical Association, 311*(5), 463–464.

Jefferson, T., Jones, M., Doshi, P., Spencer, E. A., Onakpoya, I., & Heneghan, C. J. (2014). Oseltamivir for influenza in adults and children: Systematic review of clinical study reports and summary of regulatory comments. *British Medical Journal, 348*, g2545.

Keller, A., & Bucher Della Torre, S. (2015). Sugar-sweetened beverages and obesity among children and adolescents: A review of systematic literature reviews. *Childhood Obesity, 11*(4), 338–346.

Kuhn, T. S. (1970). *The structure of scientific revolutions* (2nd ed. enlarged). Chicago, IL: The University of Chicago Press.

Kulkarni, A. V., Aziz, B., Shams, I., & Busse, J. W. (2009). Comparisons of citations in Web of Science, Scopus, and Google Scholar for articles published in general medical journals. *Journal of the American Medical Association, 302*(10), 1092–1096.

Larkin, I., & Loewenstein, G. (2017). Business model-related conflict of interests in medicine: Problems and potential solutions. *Journal of the American Medical Association, 317*(17), 1745–1746.

Lenzer, J. (2013). Half of panelists on controversial new cholesterol guideline have current or recent ties to drug manufacturers. *British Medical Journal, 347*, f6989.

Lenzer, J. (2016). When is a point of view a conflict of interest? *British Medical Journal, 355*, i6194.

Loewenstein, G., Sah, S., & Cain, D. M. (2012). The unintended consequences of conflict of interest disclosure. *Journal of the American Medical Association, 307*(7), 669–670.

Malik, V. S., Willett, W. C., & Hu, F. B. (2009). Sugar-sweetened beverages and BMI in children and adolescents: Reanalyses of a meta-analysis. *The American Journal of Clinical Nutrition, 89*(1), 438–439.

Massougbodji, J., Le, B. Y., Fratu, R., & De, W. P. (2014). Reviews examining sugar-sweetened beverages and body weight: Correlates of their quality and conclusions. *The American Journal of Clinical Nutrition, 99*(5), 1096–1104.

Mindell, J. S., Reynolds, L., Cohen, D. L., & McKee, M. (2012). All in this together: The corporate capture of public health. *British Medical Journal, 345*, e8082.

Neira, M., & de Onis, M. (2005). Preventing obesity: A public health priority in Spain. *Lancet, 365*(9468), 1386.

Neira, M., & de Onis, M. (2006). The Spanish strategy for nutrition, physical activity and the prevention of obesity. *The British Journal of Nutrition, 96*(Suppl 1), S8–S11.

Popper, K. R. (1959). *The logic of scientific discovery*. London: Hutchinson.

Raymond, N. (1998). Confirmation bias: A ubiquitous phenomenon in many guises review of general. *Psychology, 2*(2), 175–220.

Relman, A. S. (1980). The new medical-industrial complex. *The New England Journal of Medicine, 303*(17), 963–970.

Relman, A. S. (1984). Dealing with conflicts of interest. *The New England Journal of Medicine, 310*(18), 1182–1183.

Rosenbaum, L. (2015a). Understanding bias—the case for careful study. *The New England Journal of Medicine, 372*(20), 1959–1963.

Rosenbaum, L. (2015b). Conflicts of interest: Part 1: Reconnecting the dots – Reinterpreting industry-physician relations. *The New England Journal of Medicine, 372*(19), 1860–1864.

Rosenbaum, L. (2015c). Beyond moral outrage – Weighing the trade-offs of COI regulation. *The New England Journal of Medicine, 372*(21), 2064–2068.

Royo-Bordonada, M. A. (2014). The Spanish experience of public-private partnerships with the drinks and food industries. *British Medical Journal, 348*, g1189.

Royo-Bordonada, M. Á. (2016). Setting up childhood obesity policies in Europe. *The Lancet, 388*, 2475.

Royo-Bordonada, M. Á. (2017). Desafíos de las políticas nutricionales en España: de la responsabilidad individual a la responsabilidad colectiva. *La Revista Española de Nutrición Humana y Dietética, 21*(Suppl. 1), 56–58.

Royo-Bordonada, M. Á., & Román-Maestre, B. (2015). Towards public health ethics. *Public Health Reviews, 36*, 3.

Ruanpeng, D., Thongprayoon, C., Cheungpasitporn, W., & Harindhanavudhi, T. (2017). Sugar and artificially sweetened beverages linked to obesity: A systematic review and meta-analysis. *QJM, 110*(8), 513–520.

Saver, R. S. (2012). Is it really all about the money? Reconsidering non-financial interests in medical research. *The Journal of Law, Medicine & Ethics, 40*(3), 467–481.

Schopenhauer. (2014). *Parerga and paralipomena. Short philosophical essays* (Vol. I). Cambridge: Cambridge University Press.

Serra Majem, L. (2016). Conflictos de interés en la investigación en nutrición: ¿mucho ruido o muchas nueces? *Revista Española de Comunicación en Salud*, S121–S128.

Shaw, D. M., Morfeld, P., & Erren, T. C. (2016). Conflict or confluence of interest? *Journal of the American Medical Association, 315*(16), 1793.

Shawwa, K., Kallas, R., Koujanian, S., Agarwal, A., Neumann, I., Alexander, P., et al. (2016). Requirements of clinical journals for authors' disclosure of financial and non-financial conflicts of interest: A cross sectional study. *PLoS One, 11*(3), e0152301.

Stead, W. W. (2017). The complex and multifaceted aspects of conflicts of interest. *Journal of the American Medical Association, 317*(17), 1765–1767.

Steinbrook, R., Kassirer, J. P., & Angell, M. (2015). Justifying conflicts of interest in medical journals: A very bad idea. *British Medical Journal, 350*, h2942.

Steinman, M. A., Shlipak, M. G., & McPhee, S. J. (2001). Of principles and pens: Attitudes and practices of medicine housestaff toward pharmaceutical industry promotions. *The American Journal of Medicine, 110*, 551–557.

Thompson, D. F. (1993). Understanding financial conflicts of interest. *The New England Journal of Medicine, 329*(8), 573–576.

Trumbo, P. R., & Rivers, C. R. (2014). Systematic review of the evidence for an association between sugar-sweetened beverage consumption and risk of obesity. *Nutrition Reviews, 72*(9), 566–574.

Vartanian, L. R., Schwartz, M. B., & Brownell, K. D. (2007). Effects of soft drink consumption on nutrition and health: A systematic review and meta-analysis. *American Journal of Public Health, 97*(4), 667–675.

Vineis, P., & Saracci, R. (2015). Conflicts of interest matter and awareness is needed. *Journal of Epidemiology and Community Health, 69*(10), 1018–1020.

World Health Organization. (2016). *Report of the Commission on Ending Childhood Obesity*. Geneva: WHO. Accessed January 12, 2018, from http://apps.who.int/iris/bitstream/10665/204176/1/9789241510066_eng.pdf

The manufacturer's authorised representative in the EU is Springer Nature Customer Service Centre GmbH, Europaplatz 3, 69115 Heidelberg, Germany. If you have any concerns regarding our products, please contact ProductSafety@springernature.com

Printed and bound by CPI Group (UK) Ltd, Croydon, CR0 4YY
05/10/2026
02228131-0004